MANAGING HIGH GRADE DAIRY COWS IN THE TROPICS

John Moran

CSIRO
PUBLISHING

National Library of Australia Cataloguing-in-Publication entry

Moran, John, 1945–

Managing high grade dairy cows in the tropics/by John Moran.

9780643107403 (pbk.)
9780643107892 (epdf)
9780643107908 (epub)

Includes bibliographical references and index.

Dairy farming – Southeast Asia – Handbooks, manuals, etc.
Dairy farms – Southeast Asia – Management.
Farms, Small – Southeast Asia – Management.

636.21420959

Published by
CSIRO PUBLISHING
36 Gardiner Road, Clayton VIC 3168
Private Bag 10, Clayton South VIC 3169
Australia

Telephone: [+613] 9545 8555
Email: csiropublishing@csiro.au
Website: www.publishing.csiro.au

All photographs are by John Moran.

Set in 10.5/13 Adobe Minion Pro and Helvetica Neue LT Std
Edited by Peter Storer Editorial Services
Cover and text design by James Kelly
Typeset by Desktop Concepts Pty Ltd, Melbourne
Printed by Ingram Lightning Source

Jun26_RP_ILS

Foreword

Throughout the developing countries the dairy industry stands tall relative to other ruminant industries, when measured in terms of the extent of potential socio-economic impacts and contribution to nutritional and food security. One outstandingly successful example of this fact is the Anand model of India's 'Operation Flood', involving some 13 million farming families and processing about 90 million kg of milk per year mainly from dairy buffaloes. The scheme makes farmers shareholders of the whole chain of marketing and processing of milk, with resultant significant improvements to their livelihoods. Milk consumption is rapidly on the increase, and is reflected in FAO data that the annual per capita consumption of milk in the developing countries is projected to increase from 46 kg in 2002 to 67 kg in 2030.

It is appropriate therefore that most governments have given dairy development very high priority for two main reasons. Firstly, there are direct benefits to enhancing poverty-reducing impacts, improved livelihoods and the nutrition of adults and children. Secondly, considerable opportunities are created for small farmers and the landless to enter improved dairy farming, diversify agricultural production, increase income, enhance food security and achieve levels of self-sufficiency.

One direct consequence of this thrust is interest and expansion of dairy production in most countries, involving increased investments in infrastructure and germ plasm. Given the generally low quality of indigenous dairy cattle and weak breeding programs, major attention was given to the convenience of importing purebred cattle and also riverine buffaloes to boost milk production.

A historical sweep and analyses of the results of these various dairy cattle importations in Asia suggest that, in general, there have been more failures than successes. The consistent reasons were over-investments in infrastructure, poor understanding of production systems, and inefficient management and capacity for dairying. In other words, the requirements of the imported improved animals and their specific needs were not matched by adaptation to the local environmental outlays.

In many situations, the limitations and resulting unsuccessful dairy production continue to persist, often exacerbated by lack of well-informed knowledge on the subject. This particular situation has now been admirably fortified by the publication of *Managing High Grade Dairy Cows in the Tropics*. Along with three other companion books on the same subject, access to these most useful, relevant and practical sources of information should henceforth stimulate improvements and increase productivity from dairy cattle. Few scientists in Australasia have the deep knowledge and experience in practical dairy science as John Moran, and I congratulate him for his enthusiasm and tremendous efforts.

Together with the valuable sources of information, the book also serves as a training manual. I have no doubts that it will be of much interest to dairy scientists, dairy farmers, university and college students, extension personnel, farm mangers and dairy advisors in the tropics and elsewhere.

C. Devendra, PhD, DSc (Nott.), FASc
Consulting Tropical Animal Production Specialist,
30A Jalan Awan Jawa,58200 Kuala Lumpur,
Malaysia
Email: cdev@pc.jaring.my

About the author

For 30 years, Dr John Moran was an Australian senior research and advisory scientist from Victoria's Department of Primary Industries (DPIV), located in northern Victoria. Over the last 10 years, he spent half his time advising farmers in southern Australia and half his time working with dairy farmers and advisers in South and East Asia. His specialist fields include dairy production, ruminant nutrition, calf and heifer rearing, forage conservation and whole farm business management. Following his retirement from DPIV in June 2011, John formed a consulting business, Profitable Dairy Systems, located in Kyabram, Victoria.

John graduated in 1967 with a Rural Science honours degree from New England University at Armidale in NSW, followed by a Masters degree in 1969. In 1976, he obtained a Doctorate of Philosophy in beef production from the University of London, Wye College in England.

During the 1980s, John lived in Indonesia for 3 years, working in beef cattle and buffalo research. Since 1999, he has initiated and conducted training programs on small holder dairy production to farmers, advisers and policymakers in Indonesia, Malaysia, Thailand, Vietnam, China, Pakistan, Sri Lanka and East Timor.

As a result of his Asian programs, in 2005 John wrote his first dairy manual, *Tropical Dairy Farming: Feeding Management for Small Holder Dairy Farmers in the Humid Tropics.* As well as hard copy, the book was published on the internet, making it freely available to dairy stakeholders throughout the world. In 2009, John wrote his second manual on tropical dairy farming, *Business Management for Tropical Dairy Farmers.* By July 2012, the two books had received over 93 000 'hits' on the internet, indicating their relevance particularly to Asian tertiary teaching institutes and government livestock departments.

During 2009, John was commissioned by LIVECORP, Australia's coordinating agency for exporting cattle and sheep to many countries throughout the world, to develop a series of management packages for tropical and temperate dairy production technology. The background review for these packages forms the basis of this book – John's third tropical dairy manual – entitled *Managing High Grade Dairy Cows in the Tropics.*

John has published more than 200 research papers and advisory articles. He has also written several farmer manuals on dairy and beef cattle nutrition, veal production, calf and heifer rearing and silage production. The first edition of *Calf Rearing: A Guide to Rearing Calves in Australia*, published in 1993, sold more than 10 000 copies. The second edition, published in 2002, is still selling widely throughout Australia. He also published a companion book on young stock management: *Heifer Rearing: A Guide to Rearing Dairy Replacement Heifers in Australia.* His book, *Forage Conservation: Making Quality Silage and Hay in Australia*, published in 1996, is now a set text for undergraduate study in several Australian universities. His latest book, *Feedpads for Grazing Dairy Cows*, which he wrote in 2010 with his DPI colleague Scott McDonald, is also a best seller in the dairy industry.

John's initial training in a systems approach to livestock science, together with his many years working closely with dairy industries in Australia and South and East Asia stands him in great stead to write this third manual, which complement his two previous books on tropical dairy farming. *Managing High Grade Dairy Cows in the Tropics* is both a training manual for farmers, dairy advisers, technical staff and other tropical dairy specialists, plus a series of observations on how successful and profitable small holder farmers manage their day-to-day farming practices.

Other books and technical manuals by the author

Books

Calf Rearing: A Guide to Rearing Calves in Australia (1993)
Forage Conservation: Making Quality Silage and Hay in Australia (1996)
Heifer Rearing: A Guide to Rearing Dairy Replacement Heifers in Australia (with Douglas Mclean) (2001)
Calf Rearing: A Practical Guide (2002)
Tropical Dairy Farming: Feeding Management for Small Holder Dairy Farmers in the Humid Tropics (2005)
Business Management for Tropical Dairy Farmers (2009)
Feedpads for Grazing Dairy Cows (with Scott McDonald) (2010)
Fifty Years of Farmer Extension for Victoria's Dairy Industry (2011)
Rearing Young Ruminants on Milk Replacers and Starter Feeds (with Uppoor Krishnamoorthy) (2011)
Rearing Young Stock on Tropical Dairy Farms in Asia (2012)

Technical manuals

Maize for Fodder: A Guide to Growing, Conserving and Feeding Irrigated Maize in Northern Victoria (with Ken Pritchard) (1987)
Growing Calves for Pink Veal: A Guide to Rearing, Feeding and Managing Calves for Pink Veal in Australia (1990)
Growing Quality Forages for Small Holder Dairy Farms in Indonesia (2001)
Managing Dairy Farm Costs: Strategies for Dairy Farmers in Irrigated Northern Victoria (2002)
Feeding Management for Small Holder Dairy Farmers in Thailand (2002)
Improving Milk Composition Through Better Feeding Management (2003)
The Key Drivers of Good Reproductive Performance on Indonesian Dairy Farms (2006)
Value Adding Indonesia's Dairy Industry: Developing Cottage Industries in East Java (2006)

Improving Business Skills of Small Holder Dairy Farmers in Thailand (2007)
Dairy Production in Malaysia with Particular Reference to Milk Quality (2007)
Dairy Production in Indonesia with Particular Reference to Milk Quality (2007)
Developing a Post-Arrival Herd Management Protocol for Imported Australian Dairy Heifers (2007)
Managing Heat Stress and Housing for Pakistani Dairy Cows and Buffaloes (2007)
Dairy Production From Smallholder Farmers in Asia: Synopses of Workshops from 2000 to 2007 (2008)
Dairy Production in Sri Lanka: Current Situation and Future Prospects (2008)
Improving Business Skills in Vietnam's Small Holder Dairy Industry (2008)
A Guide to Better Dairy Herd Management in the Tropics (2009)
Best Management Practices for Small Holder Dairy Farmers in Asia (2010)
A Blue Print for a 50 Cow Farm in Asia (2010)
A Blue Print for a Large Scale Intensive Dairy Farm in Asia (2012)
Guide to e-modules on 'A Short Course in Tropical Dairy Farming' (2012)
Guide to e-modules on 'Feeding management on small holder dairy farms in the tropics' (2012)

Acknowledgements

Over the last 10 years, increasing numbers of Australian dairy stock have been exported to overseas markets, many of which are in South and East Asia. The rationale of such countries seeking genetically superior, but poorly adapted, dairy stock has frequently been questioned.

Unfortunately, there are increasing instances where such exercises have led to unacceptably high mortalities rates and poor performance once these imported stock have entered the milking herds, particularly those of small farmers. The major oversight by both the importers – whether private investors or government organisations – and the farmers for whom these stock are destined, is not 'preparing the environment' for the imported stock.

In collaboration with the importing country, the Australian Government has instigated strict animal selection and quarantine procedures prior to transport, while the exporting agencies are developing post-arrival protocols to reduce stresses on the newly arrived stock. The success of such programs depends largely on the awareness and ability of new owners of these stock to improve their feeding and general herd management. Unfortunately, such programs appear to be very *ad hoc*, with very variable success.

To improve the awareness, and hence the ongoing herd and farm management practices of traditional farmers in these countries, a more structured education program should be considered. Each importing country would have their specific problems facing newly arrived exotic dairy stock. Rather than producing a post-arrival protocol for each country, it is desirable to develop a generic one highlighting the common key issues. This manual highlights the important dairy herd management practices required to reduce undue stress on high genetic merit, but unadapted, dairy heifers once they arrive at their new destination.

Following the successful publication of my book *Tropical Dairy Farming* in December 2005, CSIRO Publishing asked me to write a second manual in 2009, *Business Management for Tropical Dairy Farmers*. As with my first two books, my gratitude goes out to ATSE Crawford Fund, who provided generous financial support, and to Ted Hamilton and the CSIRO Publishing team who very professionally converted my laptop computer writings to both a 'hard copy' book and a collection of PDF files for an eBook.

The commission to develop the management packages for tropical dairy production technology originated from Meat and Livestock Australia and I am very grateful for their support. I also acknowledge the team that helped produce a glossy brochure on small holder dairy farming in Indonesia: those both in Java (Tetra Pak, De Laval and the

Indonesian Association of Dairy Cooperatives) and Australia (Victorian and Queensland governments). Finally, this book was written in part while I was working full time as a dairy adviser at DPIV Dairy Extension Centre, and I thank the management team for supporting my endeavours.

Dr John Moran
Director, Profitable Dairy Systems,
24 Wilson St, Kyabram, Victoria 3620, Australia
Telephone: +61 418 379 652 (mobile phone)
Email: jbm95@hotmail.com

Acknowledgement of The Crawford Fund

The publication of this book would not have been possible without the generous assistance of The Crawford Fund whose mission is to increase Australia's engagement in international agricultural research, development and education for the benefit of developing countries and Australia.

THE CRAWFORD FUND

For a Food Secure World

Chemical warning

The registration and directions for use of chemicals can change over time. Before using a chemical or following any chemical recommendations, the user should **always** check the uses prescribed on the label of the product to be used. If the product has not been recently produced, users should contact the place of purchase, or their local reseller, to check that the product and its uses are still registered. Users should note that the currently registered label should **always** be used.

Contents

1

Introduction

This chapter presents an outline of the manual, which highlights the basic components of successful small holder dairy farming in tropical climates.

The main points in this chapter

- This is the third book I have written on tropical dairy farming, with the previous two concentrating on feeding management and on-farm business management of small holder farms.
- This book deals specifically with overcoming the many problems of poor adaptation of exotic high grade dairy stock to the stresses of tropical climates and small holder herd management.
- Small holder dairy farmers (with herds up to 20 milking cows plus replacement heifers) are generally competitive and sustainable.
- Dairy development is associated with technical changes to improve milk yield per cow.
- Most countries have development programs involving importing high genetic merit dairy stock, usually Friesians.
- Dairy production technology can be broken down to nine links in a supply chain on any dairy farm, no matter its size or location.

This book is a companion to two previous books I have written on small holder dairy (SHD) farming in the tropics. The first book, *Tropical Dairy Farming* (Moran 2005), details the production technology of SHD farming, with emphasis on nutrition and feeding management. The second book, *Business Management for Tropical Dairy Farmers* (Moran 2009a), discusses the farm business management (FBM) skills required to ensure such systems can remain financially sustainable. This third book, *Managing High Grade Dairy Cows in the Tropics*, deals specifically with a major problem encountered by many tropical dairy farmers: namely the poor performance of exotic, high grade (that is, high genetic merit) dairy cows when exported from their country of origin to a new, more stressful environment.

The first book addresses the management of the farm's natural and biological resources to produce quality milk – the climatic environment, soils, forages, concentrates and the livestock. The second book concentrates more on the human side of SHD farming, namely the farming family and their support structures, which include village communities, cooperatives, marketing and government agencies and other service providers and, importantly, the consumers. This third book incorporates a checklist to assess current farm management practices with an example framework for grading the suitability of individual farmers to receive high-quality dairy stock.

Over 75% of the world's poor people (i.e. 2.6 billion people) live in rural areas, with many of these dependent on farming for their livelihood. SHD farming is an enormous global industry, with 12–15% of the world's population (i.e. 900 million people) living on 150 million SHD farms. In the Indian subcontinent alone, there are 75 million SHD farms in India and 15 million SHD farms in Pakistan. Herd sizes on these farms are only 2.4 cows per farm, with each farm producing only 11 kg/day of milk. However, milk production supports a very large post-farm gate workforce, because every million kilograms of milk creates 200 jobs, compared with only five jobs in developed countries.

During the last three to four decades, governments throughout Asia have established SHD farming as part of their social welfare and rural development schemes, to provide a regular cash flow for poorly resourced and often landless farmers. Now these have become accepted rural industries. The need is for a more business-minded approach to management decisions on each farm. Dairy farmers across the world, even the small holder mixed farmer with only one or two cows, milk cows to make money. As the dairy value chain becomes more liberalised, and farmers become more exposed to the pressures of global markets, their daily farming decisions must become more based on changes to their farm profitability. This and the previous manuals describe the technical principles, practical processes and business decisions required to optimise their farming operations both in terms of profit and long-term sustainability.

1.1 Who are the tropical small holder dairy farmers?

Geographers categorise the humid (or rainy) tropics as areas with at most one or two dry months and no winter, with the coolest month above 18°C mean temperature. Other tropical zones are:

- **wet and dry tropics**, which have a well-developed dry season, with one or two rainy seasons
- **semiarid tropics**, with light rainfall and high evaporation
- **hot arid tropics**, with negligible rainfall and high evaporation.

My first book, *Tropical Dairy Farming* (Moran 2005) limited its scope to the humid tropics of South-East (SE) Asia. As with my second book, *Business Management for Tropical Dairy Farmers*, this book extends its scope to all the above zones within tropical Asia, although dairying is more likely to be restricted to just the humid and the wet and dry tropics. In addition, it covers both South and East Asia, not just SE Asia, as in *Tropical Dairy Farming*. Tropical Africa and Central America are not discussed in any great detail.

There are often arguments as to what constitutes small holder (as against large-scale) dairying. This manual uses the following descriptors of dairy farms:

- **Small holder:** up to 20 milking cows plus replacement heifers
- **Semi-commercial:** 20 to 50 milking cows plus replacement heifers
- **Commercial:** more than 50 milking cows plus replacement heifers.

SHD farmers are generally competitive and are likely to endure for many years to come, particularly where the opportunity costs of family labour and wages remain low. In addition, dairying is a viable enterprise even among the landless and socially marginalised groups.

Policymakers should resist the all too common assumption that development efforts should move from small holders towards supporting larger scale, 'more efficient' milk producers to meet growing consumer demand. Instead, that growing demand should be used as a stimulus to help continue and sustain SHD enterprises, particularly when they face increasing barriers to participate in value chain markets.

1.2 Dairy development and farm technology

Dairy development is generally associated with technical changes to improve milk yield per cow. However, it should be noted that:

- The use of exotic cattle is a rapid and potentially sustainable path to higher productivity, even for small-scale resource-poor farmers and in warm, semi-arid or humid climates. However, there have been many repeated failures of such schemes for obvious, but often ignored, reasons.
- National and local breeding strategies need to address the realities of climate and disease risk to increase the likelihood of successful crossbreeding programs.
- Fodder technology should be an integral part of any dairy development program, particularly if it incorporates importation of high genetic merit stock.

After several decades of dairy development in many Asian countries, average milk yields per cow per day still range between 8 and 10 kg, compared with average yields of 20–30 kg in developed countries. In addition, calving intervals of dairy cows on Asian SHD farms is commonly as long as 16–20 months, when it could be reduced to 14–15 months. This clearly shows their low levels of farm productivity. Some technical solutions are available but they must be carefully selected so they will be suitable for small farmers and their socio-economic conditions. This means that scientists and extension worker must be able to understand factors influencing the acceptance of technology by SHD farmers.

Granted, the genetic potential of local cattle to produce milk can become a constraint as the quality and quantity of farm inputs increase, but the introduction of new improved breeds must be accompanied by other investments in feeding, health, hygiene and housing, lest the system fail.

Scientific knowledge alone cannot solve small-scale farm problems. As well as technological innovations to improve farm performance, there is considerable knowledge on dairy herd management that is not being readily adopted by small holder farmers. In all too many cases, the flow of such information does not reach the individual farmer even though it has been shown to be applicable to many farming situations. This then is the rationale behind this manual in that it presents current farm practices that are already being routinely used by many successful small holder farmers throughout the tropical dairy industries.

1.3 Importing high genetic merit stock onto small holder dairy farms

Most countries with SHD industries have development programs involving increasing cow numbers and genetic quality through importing dairy stock, usually Friesians. This is because their rate of natural multiplication of their national dairy herds is too slow to supply the stock to satisfy the increasing demand of milk and other dairy products.

The major oversight by both the importers – whether private investors or government organisations – and the farmers for whom these stock are destined, is not 'preparing the environment' for the imported stock. The greatest shortfalls are:

- lack of knowledge of the quality of local feedstuffs, particularly forages
- lack of understanding of the cows' nutrient requirements for acceptable performance, to reduce stress
- low skill levels of local labour to handle the high level of technology in genetics of imported stock
- poor sanitation practices for manure disposal, fly control and drying of all floor surfaces
- lack of sufficient quarantine to minimise spread of disease while heifers are still susceptible
- lack of knowledge and management skills to address problems during parturition
- difficulty in supplying an optimum diet during early lactation to ensure limited live weight loss, hence short lactation anoestrus
- minimising environmental stress during early lactation so newly calved heifers will cycle normally after 2 months.

Other factors to consider include:

- selecting the most appropriate heifers prior to transport
- providing good calf and heifer rearing management so that calves from imported heifers are well grown and have the opportunity to express their true genetic merit when milking.

1.3.1 Genetic merit of imported stock

The decision on the most appropriate type of stock to import should be seriously considered. Selection of high genetic merit heifers or cows to import will force farmers to improve their feeding management to provide sufficient energy to allow these animals to better express their superior genetic merit. This is very important in early lactation, when such stock are expected to cycle as well as produce high levels of milk.

Cows can produce the same milk yield in early lactation with whole body energy balances ranging from +4 to –25 MJ/day (see Chapter 5). This is the energetic equivalent of 0.8 kg milk to build body reserves or to lose the energy derived from losing 0.8 kg/day of body weight. Clearly cows that are genetically 'programmed' to lose excess weight in early lactation are less desirable in small holder systems where feed shortages are all too common.

Perhaps it would be wiser to import stock of lower genetic merit, and then feed them to produce less milk, but at least improve their chances of getting back in calf within 100 days post-partum. The best farmers in Australia can achieve 58% 100-day-in-calf rates (Morton *et al.* 2003), compared with only 40% in a well-managed farm in Vietnam (Moran and Tranter 2004).

It is not uncommon for Asian importers of Friesian heifers from Australia to request '5000 litre heifers'; that is, heifers likely to produce 5000 L during their first lactation on their home farm. It is highly likely that, once in their new farm in Asia, such animals would only be capable of producing 2000 or 3000 L during their first (and even later) lactation given the existing feeding management. If such animals, managed to produce 25 L/day in peak lactation (as a starting point to achieve their 5000 L first lactation yield), their daily energy deficit would be highly likely to impair their fertility.

1.3.2 Importing young heifers

The high cost of importing pregnant heifers makes alternative methods of increasing cow numbers feasible. For example, consideration could be given to purchasing heifer calves, and growing them out for say 6–12 months in their home country prior to their importation. Granted, this will only supply one animal, compared with the heifer plus embryo when importing pregnant heifers, and delay initiating income, but this will allow an additional 12 months for adaptation to the new environment before calving.

Boat carrying Australian dairy heifers coming into the Kuala Lumpur offloading port (Malaysia).

Preparing for the unloading of stock (Malaysia).

Such adaptation includes adjustment to the climate, feeds and management regime, and developing resistance to local disease organisms. Over the short term, importing pregnant heifers may give higher returns, but in the long term, importing younger heifers can be more economic (McDowell 1994).

This discussion does not even consider the greater susceptibility of high genetic merit cows to their physical environment. High milk yields, through increased feed intake, generate more internal heat, thus requiring a less stressful environment to dissipate such heat. Therefore, unless such cows are allowed to regain their normal heat balance, their appetite will fall, so they will eat less. The only way they can produce milk close to their desired level is by partitioning body reserves towards milk synthesis, thereby increasing energy deficits through greater live weight losses.

It is then likely that these animals will be unable to produce their target milk yields and will not easily get back in calf. Such animals are generally more susceptible to other constraints of the tropical environmental, such as parasites and disease. This further increases their likelihood of being culled as non-productive animals. Even if their high purchase cost and their 'status' as exotic cows reduces pressures to cull them, they will become poorer investments compared with locally adapted, and less potentially productive stock.

1.3.3 The renewed relevance of embryo transfer technology

The practice of multiple ovulation and embryo transfer (MOET) is only slowly being accepted in Western dairy industries as another method to more rapidly improve genetic merit, albeit mainly in the pedigree dairy industry. The general consensus is that MOET is not yet a sound economic approach for widespread use in dairy operations whose entire or main business is selling milk. This was first mentioned by McDowell in 1994 and is equally valid in 2012.

However, for large-scale multiplication of national herds – where genetic progress is less of an issue and where, in their new country as lactating cows, imported heifers cost two to three times more than their original purchase cost – MOET may have an economic role. Countries such as China are currently assessing its relevance to national industry development.

1.4 Outline of the manual

This book is written primarily for the stakeholders of SHD production in the tropics. Small holders are the major suppliers of milk in the tropics. However, numerous larger farms with many hundreds of milking cows, using intensive feedlot or less intensive grazing systems, have been established throughout South and East Asia in recent years to satisfy the increasing demand for fresh milk. These farmers and their advisers will also gain much from this manual. In addition, the book provides relevant key information to research scientists on aspects of tropical dairy production and business management, such as forage production, and herd and feeding management. Policymakers and senior managerial personnel would also benefit from reading selected chapters.

Most tropical countries have proactive programs to increase local supplies of milk, which require increasing numbers of well-trained workers to service the dairy industry. Consequently, educators from agricultural schools, universities and technical colleges need to be kept abreast of the latest technical developments and applications in dairy farming.

Chapter 2 covers the pre-departure planning and management of stock on arrival at their new destination. Chapter 3 reviews the soil and forage management on the new farm to ensure adequate year-round supplies of quality forage: the essence of any profitable tropical dairy farming operation. This includes a section on silage to make the best use of any excess forage production. Chapter 4 discusses the management of young stock, covering both the pre-calving period of imported heifers and the early milk feeding period of their progeny. Scours, a major calf rearing problem on many small holder farms, is discussed in detail.

Chapters 5 and 6 discuss improved nutrition and feeding management on tropical SHD farms, firstly highlighting the key principles of dairy nutrition (Chapter 5) then some of the more successful practices (Chapter 6). Chapter 6 also provides insights into feeding problems and unbalanced diets, such as lactic acidosis.

Disease prevention and control are covered in Chapter 7, with particular emphasis on two common problems on SHD farms, namely lameness and mastitis. Chapter 8 deals with improved reproductive management, an all too common problem with high-yielding dairy cows. Chapter 9 discusses some of the farm decisions regarding genetic improvement of the dairy herd, while Chapter 10 addresses the management of any stock surplus to the milking herd. Chapter 11 highlights an ongoing issue with exporting countries, namely the welfare of confined animals under traditional farm management.

The principles and practices of overcoming heat stress are discussed in Chapter 12 and more specifically in Chapter 13, with recommended housing systems for tropical SHD farms. Milk harvesting and milking hygiene are reviewed in Chapter 14, while Chapter 15 describes an Indonesian case study of value adding to milk.

The business management of SHD farming is reviewed in Chapter 16, while Chapter 17 describes the planning and implementation of farmer workshops on improved herd management. Chapter 18 incorporates a checklist to assess current farm management practices with an example framework for grading the suitability of individual farmers to receive high-quality dairy stock. Chapter 19 is a series of photographs depicting tips and traps in managing high grade dairy stock around Asia.

Full publication details of all sources of information are presented in 'References and further reading'.

Every profession has its jargon, or words developed specifically for that profession, and agriculture is no exception. There are some very specific terms and acronyms that are routinely used by farm management economists and consultants. These are explained in the Glossary and when they are first used in this book.

Appendices are included to facilitate sourcing specific information and learning the skills of ration formulation. Appendix 1 presents the Temperature Humidity Index: the universal method of quantifying heat stress in dairy stock. Appendix 2 provides conversion factors to the standard metric system from a wide variety of systems used for describing weights and measures. Appendix 3 presents typical expectation and evaluation forms developed for farmer workshops on improved herd management.

In the process of developing a series of workshops on dairy production technology for Indonesian small holder farmers, I prepared a summary on tips for proper managing of dairy cows on Indonesian small holder farmers. Appendix 4 presents this four-page booklet, which highlights the key features of good feeding, herd and milking management on profitable farms throughout Asia.

Finally, for ease of finding specific information, the Index lists all the key topics covered in the book and their relevant page numbers.

1.5 The role of the manual in training programs

This manual is multipurpose in that it forms the basis of structured training programs in small holder dairying for advisers and educators (for farmer training organisations, agricultural high schools and universities), while also providing background information to researchers and policymakers in tropical dairy industries. Unlike my previous manuals, a series of training programs were developed alongside this book and are being used for adviser and farmer workshops in the SHD industries that import Australian dairy stock (MLA 2009). This manual has been written as a reference text for these training programs.

These chapters are written to be understood by advisers and tertiary students. Because the trainers must ensure that other target audiences can comprehend their course material, they should select the most relevant sections to incorporate into basic programs for farmers. Because each chapter has been written as a 'stand alone' document that can be individually downloaded from the Internet, there is some repetition, but this has been kept to a minimum.

1.6 The key task areas in any dairy production system

Dairy production technology can be broken down into nine key task areas (see Figure 1.1), which can be considered as steps in the supply chain of profitable dairy farming, whether a SHD or large holder. Just as any chain is only as strong as its weakest link, each step in this supply chain must be properly managed. Weakening any one link through poor decision making can have severe ramifications on overall farm performance, and hence profits. In chronological order of their role in ensuring a profitable dairy enterprise, the 'links' in Figure 1.1 and the relevant chapters in this manual are as follows:

1. Chapter 3. Soils and forage management, to optimise forage agronomy and fodder conservation.
2. Chapter 4. Young stock management, to generate productive milking cows.
3. Chapter 6. Feeding management, to optimise cow performance through adequate supplies of feed nutrients.
4. Chapter 7. Disease prevention and control, to overcome the limits imposed by microbial and invertebrate pests.
5. Chapter 8. Reproductive management, to ensure herds can replace themselves in future generations.
6. Chapter 9. Genetics, to maintain an acceptable rate of genetic improvement for each generation.
7. Chapter 12. Environmental management, to limit the constraints of climate on cow and heifer performance.
8. Chapter 14. Milk harvesting and hygiene, to maximise milk quality pre- and post-farm gate.
9. Chapter 15. Adding value to milk, to improve unit returns for raw milk.

This approach then systematically addresses the key aspects of herd and feed management that require attention when including high genetic merit stock in the milking herd. Additional chapters are included in this manual to aid understanding of some of the technical aspects of these basic components. These are:

- Chapter 2. Planning the consignment and management of the imported stock immediately on arrival.
- Chapter 6. Principles of dairy nutrition, as background information on feeding management of the milking herd.
- Chapters 10 and 11. Managing stock surplus to the milking herd and stock welfare, to complement other chapters on herd management.
- Chapter 13. Housing systems, to provide more specific details on environmental management.
- Chapter 16. Business management of small holder dairy farms, to assist with implementing farm practices to optimise farm profits and long-term sustainability.

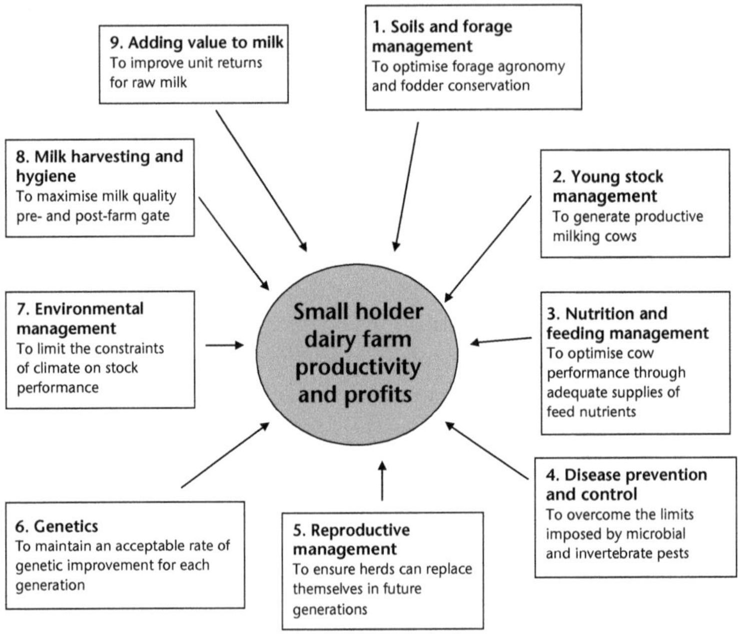

Figure 1.1. The nine links in the supply chain for a profitable dairy enterprise

- Chapter 17. Conducting farmer workshops on improved herd management, so farmers can plan and implement relevant strategies.
- Chapter 18. Assessing current farm practices, to highlight, and hence tackle, examples of poor herd and farm management.
- Chapter 19. Tips and traps in managing high-quality dairy stock: a series of photographs depicting examples of poor and good herd management.
- Appendix 4. A four-page brochure on 'Tips for proper handling of dairy cows on Indonesian smallholder farms'. This brochure highlights the key features of good feeding, herd and milking management on profitable farms throughout Asia.

To understand more fully the key factors influencing these basic steps of dairy production technology, they have been highlighted in diagrams at the beginning of each chapter. This will provide extension workers and other dairy specialists with a checklist when using this manual to develop farmer training workshops.

As previously mentioned, this is the third manual specifically written on tropical SHD farming. There is some inevitable overlap in topics covered within these three books, so some repetition is inevitable because not all readers of this third manual will be familiar with the technical aspects covered in the first two manuals. This is particularly the case with nutrition of the milking herd, because it is important to quantify the intakes of feed nutrients to better understand any production responses to improved feeding management. Before nutrients can be quantified, they must first be defined in terms of energy, protein and fibre intake.

2

Pre-departure planning and management of stock on arrival

This chapter covers planning the consignment of imported dairy stock.

The main points in this chapter

- What the customer wants is not always what the customer needs. Furthermore, what the customer needs may even differ from what the customer eventually gets.
- The process of importing stock can be best described through the eight steps of the livestock export chain.
- The exporting agent, the Australian Government, and importing governments all have obligations to ensure the live export process is correctly undertaken.
- There are a series of criteria for rejecting dairy stock for live export and these are summarised in this chapter.
- Transportation introduces many stresses to the stock, requiring a period of recuperation that coincides with any post-arrival veterinary healthy checks.
- Immediate post-arrival management can have long-term effects on the survivability, wellbeing, and hence the subsequent performance of these dairy stock.

This chapter discusses some of the key issues to be considered by the agents representing the country importing high genetic merit dairy stock. Details refer specifically to agents importing stock from Australia, but the principles should not greatly differ to those of other exporting countries.

2.1 Satisfying customer demands

The importing country makes a valued judgement as to the type of dairy stock it needs to purchase. However, it must be emphasised that what the customer **wants** is not always what the customer **needs**. Furthermore, what the customer needs may even differ from what the customer actually **gets**.

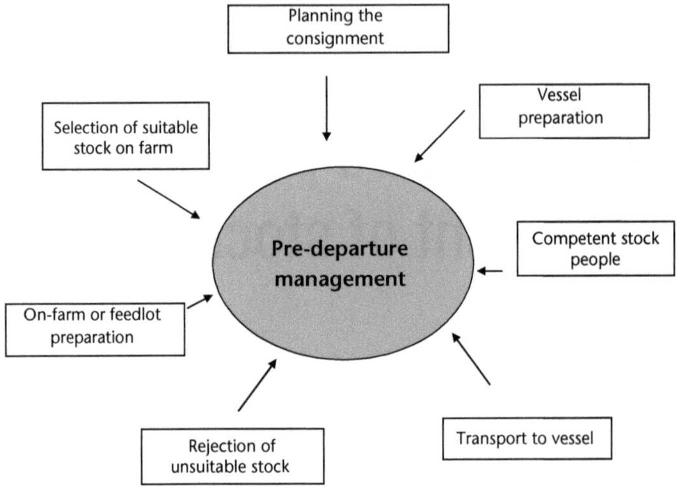

Figure 2.1. The key factors to consider when planning pre-department management of export dairy stock

2.1.1 What the customer wants

Generally speaking, the importer wants, firstly, to increase herd numbers in the region being serviced and, secondly, to improve the genetic merit of the milking cows in that region. The first can be achieved by importing any type of dairy animal that will produce milk and reproduce under the existing environmental and production constraints.

Depending on the market demand for the raw product, the imported stock can produce:

1. relatively low quantities of high solids milk, if it is destined for industrial processing to yield dairy products such as milk powder
2. higher yields of milk with lower milk solids, if it destined for consumption within a few days or even if it is to be processed into long-shelf-life products to which imported milk powder can be added.

Milking cows must also routinely produce calves, either as herd replacements or for sale, plus have a disposal value as a cull cow. This can influence the preferred breed because purebred and crossbred Friesians have a higher value for dairy beef than Jerseys or Zebus.

The purchaser may have a set of specifications, some of which are poorly related to production, such as coat colour (e.g. four white feet and white tip of tail) and udder shape (particularly for cows being milked by hand rather than machines). Large body size should be less of a selection criterion with poorer feeding management and may even impair performance and longevity in some small holder milking herds. It should be explained to the purchaser which of these selection criteria are the most relevant to the farming system to which the stock will eventually be subjected.

Some contracts may require printed proof of the cow's ancestry, such as the pedigree of its Friesian sire and proof of some stipulated minimum milk yield of its dam. Because many of the farmers supplying export stock do not record such data (whose value to the importing country must frequently be questioned), this can limit the number of stock acceptable to such purchasers.

Rather than seeking pregnant heifers, some importers prefer younger virgin heifers, with minimum live weights, such as 9–16-month-old Friesian heifers weighing from 200 to 360 kg.

2.1.2 What the customer needs

The purchaser usually requires an 'all round performer'. Initially the imported stock may be well managed, but eventually their progeny will enter the general dairy population hence be subjected to more normal farming systems. The ideal imported heifer and her progeny should then be able to perform equally well, compared with local stock, in both well and poorly managed systems. Therefore heifers from 'average' rather than 'superior' dams may be better suited to purchaser requirements.

There is always debate regarding whether it is better to source genotypes to suit existing systems rather than try and 'fit the system' around the genotype. The challenge of the importer is to match the genetic merit of the stock to those farms most likely to benefit from it.

New tropical dairy breeds such as Australian Friesian Sahiwal (AFS) and Australian Milking Zebus (AMZ) appear to be better suited to the humid tropics than the purebred Friesian. Jerseys are a more tropically adapted breed than Friesians, with their smaller body size, black skin, higher density of sweat glands and lower internal body heat production, because of their lower productivity. As well as their suitability, there is increasing interest in exporting countries, such as Australia, in using some of the production attributes of the crossbred Jersey × Friesian. In fact, these animals may have potential as export heifers, at least until more AFS become available for sale. Brown Swiss is another breed considered to have tropical adaptation features.

2.1.3 What the customer gets

There is increasing concern in many developed temperate dairy industries that current levels of heifer attrition, through export sales, is eroding the genetic quality of their dairy stock. Replacement heifers are required firstly, to expand these local industries and secondly, to allow adequate culling of inferior stock. Such concerns increase sale prices for export heifers, impacting on supplies of suitable stock to meet export demands.

Unlike in other countries, many Australian dairy farmers do not use Friesian bulls or semen over their replacement heifers, often preferring Jerseys or Angus because of small Friesian heifer size at mating. Furthermore, such crossbred progeny have high sale value as week-old calves. The use of Jersey sires is likely to be greater, with current interests in crossbred milking cows, thus reducing the availability of purebred Friesians, the genotype of choice for export.

In recent years, sourcing stock for export has been made easier as farmers offload their excess replacements during drought years or when leaving the industry. However, such supplies are abnormal compared with more normal years, when farmers require 25–30% of their heifers as herd replacements. How long the current rate of supply of dairy heifers will be available to meet the increasing demands for export must also be addressed.

2.2 Australia's livestock export chain

To minimise the chances of adverse animal health and welfare during the live export process, the correct approach to stock management should be based on the whole of chain risk (DAFF 2006). This is important to ensure that all critical risks are identified, their potential impacts analysed and the most appropriate management procedures developed. This live export chain can be broken down to the following eight steps:

1. **Planning the consignment.** This covers the entire process, from preparation and sourcing the stock until they have been unloaded in the importing country. It also includes contingency plans in case of unexpected threats to animal health and welfare.
2. **Sourcing and on-farm preparation.** All stock must meet Australian and the importing country's health, welfare and commercial requirements and this may or may not require specific on-farm preparation.
3. **Land transport.** This involves moving all the stock from their property of origin to the pre-shipping location.
4. **Pre-embarkation assembly.** This takes into account the moving of stock from the pre-shipping location to the actual vessel for export.
5. **Vessel preparation and loading.** This includes the selection and preparation of a suitable vessel to transport stock overseas. Loading includes an inspection of the stock for health and fitness to travel at the port, before they are loaded onto the vessel.
6. **Sea/air voyage.** This covers on-board management during the voyage.
7. **Disembarkation.** Following arrival at the destination and acceptance of the stock for unloading, they are unloaded, after which the Australian Government's jurisdiction over the consignment is complete.
8. **Post-disembarkation.** The health and welfare now become the responsibility of the importer, under the authority of the importing country.

The exporter's responsibility covers the sourcing, land transportation, treatment and inspection before export, and specific plans to manage stock health and welfare during the journey, whether by sea or air.

All livestock for export must meet the Australian animal health and welfare standards, which are consistent with good animal husbandry practices. The importing country determines animal health and other requirements for imported stock, which it must advise the Australian Government from time to time. The exporter is responsible to the Australian Government for compliance with these requirements. AQIS (Australian Quarantine and Inspection Service) provides a health certificate and an export permit when it is satisfied that all stock for export meet the requirements of the importing country. The guiding principle is that livestock export consignments from Australia meet the requirements of the national animal health and welfare system and importing country requirements.

The exporter must source suitable stock that meet consignment specifications, such as the number of stock, their species, class, condition, and animal health and welfare status. The exporter must also ensure that sufficient livestock services are maintained throughout the voyage and management is adequate to maintain animal health and welfare, engaging accredited stock people and, when required, accredited veterinarians. The exporter is responsible for loading the vessel in a manner that prevents injury and minimises stress, that stocking densities are acceptable and there are adequate provisions of feed, water and veterinary supplies on the vessel. The Master of the vessel assumes overall responsibility for the management and care of the stock during transport on the vessel.

2.3 Rejection of dairy stock for live export

Stock selected for export must be fit to travel. They must meet relevant state animal health and welfare standards and national codes of practice, as well as importing country requirements. They must be identified to their property of origin, be accompanied by correctly completed and signed declarations and be individually identified where testing is required during preparation.

Bos taurus cattle (which include Friesians and Jerseys) bred in an area of Australia south of latitude 26 degrees can not be sourced for export to the Middle East between May and October each year unless an agreed heat-stress risk assessment has been completed, predicting less than a 2% risk of 5% mortality. Lactating animals with young at foot can be transported only by air.

There are various standards (DAFF 2006) for rejection of cattle and those relevant to dairy stock are as follows:

- lactating animals
- if pregnancy status is not confirmed as appropriate for the journey, or they are more than 190 days pregnant at the scheduled date of departure. For air transport, breeding stock can be up to 250 days pregnant
- emaciated or overfat stock (with a score of less than 2, or 3 for pregnant stock, and more than 6) on a scale of 1 to 7 for body condition points score
- animal with anorexia or inappetence
- uncoordinated, collapsed, weak, unwell, lethargic or dehydrated stock or those suffering ill-thrift
- stock suffering from lameness, abnormal gait, abnormal soft tissue or bony swellings
- stock with dysentery or profuse diarrhoea, or bloat
- stock showing nervous symptoms, or abnormal or aggressive behaviour
- stock with generalised skin diseases
- stock showing visible external parasites or significant lacerations
- stock with discharging wounds or abscesses or blood/discharge from the vulva
- stock blindness in one or both eyes, or with excessive salivation, nasal discharge or coughing
- stock with untipped sharp horns or horns longer than 12 cm
- herds with a large disparity in size or age (these must be redrafted)
- herds with unusual mortalities during pre-export isolation

- calves weaned less than 14 days before being sourced for export
- stock weighing less than 200 kg or more than 650 kg
- stock that have become sick or injured during on-farm preparation
- pregnant stock cannot be treated with a prostaglandin drug within 14 days of export, and not during 60 days before export unless identified as being non-pregnant or less than 90 days pregnant.

2.4 Minimising the stress of transportation

Exportation of cattle involves lengthy periods of transportation. This can be by land, sea and/or air. Cattle are often required to be transported by land over relatively long distances, both within Australia, from their property of origin to reach ports prior to shipping, and on arrival at their country of destination, to travel to their new home farm. Sea transport can involve many days or even several weeks. Air transport may be relatively short, but the associated land transport to and from the airport can be lengthy.

There is increasing recognition of the need to minimise stress in livestock, both as a result of public concern for their welfare, and any adverse effect on their subsequent performance.

Animals being transported are subject to a number of stress factors throughout the journey, including handling, loading, transporting, mixing of unfamiliar animals, unloading and total time without water or food. These risk factors can be cumulative and apply across all phases of transport, from assembly and handling before the journey, to unloading at the destination.

Stress occurs when stock are exposed to changes in their normal environment. The effects of stressors are cumulative when they operate in combination. An animal already burdened with one stressor is less able to adapt to another. Therefore every effort should be made to minimise the degree of each stress so they can cope with additional ones. Table 2.1 indicates some of the main stressors likely to arise during the export process and their effects.

Animals show behavioural changes when adapting to changes in their environment. Measures of animal behaviour will provide the best overall measure of animal discomfort.

2.4.1 Clinical indicators of transport stress

Transport-related stress can be measured by the occurrence of a range of clinical syndromes. These include:

- loss of live weight
- dehydration
- decreased feed intake
- traumatic lesion such as bruising
- decreased disease resistance.

The extent to which these occur depends on a number of other interrelated factors, including the environment, transport conditions and animal factors such as nutritional

Table 2.1. Stressors in livestock transport

Stress	Stressor	Effect
Behavioural	Novelty, restraint, noise	Fear
	Mixing, overcrowding	Aggressive interaction
Nutritional	Fasting	Dehydration and hunger
Physical	Mixing, overcrowding, road conditions	Bruising and injury
	Weather extremes	Heat stress, cold stress
	Dust	Respiratory disease
Infectious	Exposure	Respiratory disease

status and the presence of infectious diseases. Road transport conditions include total transit time, overcrowding, driving care and road conditions. Sea transport is probably the most stressful because of its duration and dependence of adequate feeding management, good stock handling and exposure to disease agents while on board. Air transport also has its specific stressors, such as lack of space in crates and the requirements for adequate ventilation for many hours.

Loss of body weight is commonly found in all transported cattle. Most of this loss occurs in the first few days of transport due to periods of food and water deprivation. The combination of food and water deprivation with road and sea transport compounds the weight loss, which is due mostly to loss of body fluids and gut contents rather than body tissue.

Dehydration is the principal nutritional stress encountered and the main cause of live weight loss during transport. Water accounts for most of the weight loss and the longer the travel time, the more weight loss occurs.

Stress suppresses the immune system and therefore increases the likelihood of disease. Diseases that can be triggered by transport and environmental changes from extensive to intensive husbandry conditions include bovine respiratory disease, which is the most common disease in cattle upon introduction to feedlots such as during export transport, in quarantine yards either prior to shipping or on arrival at the new home port.

2.4.2 Management of transport stress

Management strategies for dealing with transport stress include pre-conditioning, rest periods during and after transport and the use of electrolyte solutions.

Electrolyte solutions have been used to maintain the acid–base balance of stock. However, recent research with Zebu steers subjected to land transport for 48 hr found that these animals were able to maintain their acid–base balance within normal values without the use of electrolyte solutions. Consequently, the researchers concluded that offering electrolyte solutions to dehydrated, transported, nutrient-deprived and stressed Zebu cattle was unlikely to resolve the physiological stressors any more efficiently than water alone.

Until now, hydration strategies involved with transportation have relied on the replacement of lost body water and electrolytes following the journey, or after their welfare has been compromised. Boosting body water prior to transport is only temporary because the kidney rapidly excretes any excess fluid. Electrolyte solutions fed post-transportation provide little benefit in correcting the acid–base balance, compared with water alone.

However, there are novel non-electrolyte supplements available that, when provided prior to transport, are effective in improving energy balance and minimising spare muscle protein degradation through its effect on electrolytes, glucose and the acid–base balance. Therefore treatment with these supplements achieves the desired hyper-hydration, to decrease the energy deficit and confers a positive effect on the animal's immune system. Further research is under way to better understand the effects of these supplements.

Heat stress during transport (such as on ships) can be alleviated by wetting stock, provided there is sufficient air movement around the stock. Water should be applied to the head and back of cattle, and enough water needs to be applied so that it starts to run off their backs and down their sides. Do not apply cold water (less than 25°C) to the head of cattle with severe heat stress because the cold shock may kill them. Under moderate heat stress, short duration wettings of less than 1 min per six head of cattle can be effective for up to 24 hr.

2.5 Post-arrival management

Following the journey from their country of origin, all stock need time and a suitable location to recover. During this time, they may be subjected to further veterinary attention to comply with the requirements of the importing country. The period of quarantine and the range of health checks and vaccinations will differ, depending on the veterinary regulations of each purchasing country.

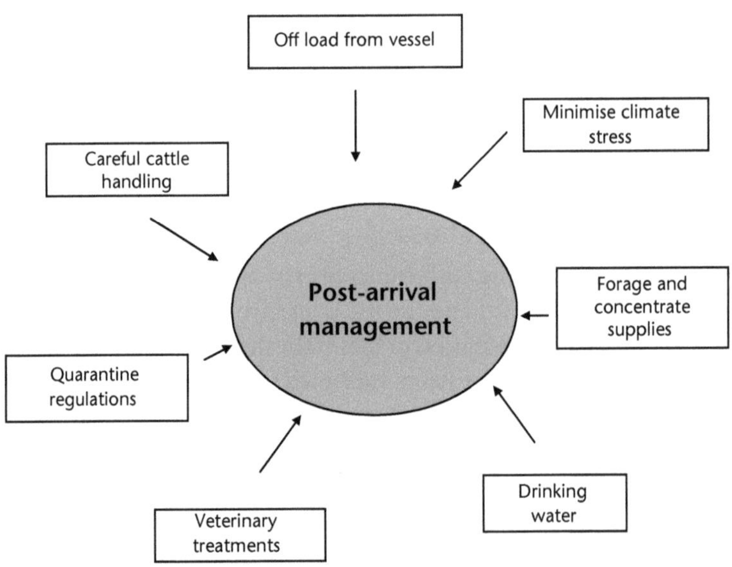

Figure 2.2. The key factors to consider when preparing for immediate post-arrival management of export dairy stock

Imported dairy heifers would not perform well on this small holder farm (Indonesia).

This small holder farmer would be unlikely to manage imported dairy heifers to their potential (Sri Lanka).

The quality of the immediate post-arrival management can have a dramatic effect on the subsequent survival, health and performance of the imported stock once they reach their new home farms. The emphasis of all such management should be on minimising the adverse effects of the accumulated old stressors and the impact of any new ones on their wellbeing. A well constructed set of cattle yards, race and crush will facilitate veterinary inspections. Reducing the adverse effects of climate stress will require shade and adequate resting space. Sufficient clean drinking water and quality feed will allow the stock to recuperate from the completed leg of their journey. Competent stock people are essential because it is likely the imported stock will not have had the exposure to close human contact as other stock on their new home farm and, in addition, their temperament is likely to still be upset from the trauma of their journey. The quarantine location is likely to be very different from the stock's normal environment, so recuperation could be quite slow.

Following their quarantine period on arrival, further road transport is likely for them to reach their new home farm. The conditions of transportation could add an additional stress to their metabolism, so it is important that they be handled quietly and be allowed to rest comfortably during early post-arrival, prior to this final journey.

3

Soil and forage management on the new home farm

This chapter discusses the key principles to good forage agronomy.

The main points in this chapter

- The basis of economic dairy farming is producing and using quality forage.
- The four basic principles of producing quality forage are:
  1. Selecting the most appropriate forage species for the region.
  2. Preparing the forage production area for sowing.
  3. Managing the agronomy of the crop.
  4. Harvesting the crop at the best stage of maturity for maximum nutritive value.
- The key to good forage agronomy is providing sufficient soil nutrients to overcome nutrient constraints to plant growth.
- Forage crops should be harvested to optimise the supply of the key nutrients for milk production, namely energy and protein.
- There are 10 key steps that should be followed to make quality silage. There are no short cuts!
- Silage also allows the long-term storage of a variety of wet agro-industrial by-products.

Because this is a manual about dairy stock management, this chapter is likely to be of less interest to the reader than other chapters. However, it is essential to be aware of the basics of growing a year-round supply of quality forage to ensure the dairy herd has a regular feed supply with which to produce the farm's livestock products.

Milking cows require a year-round supply of feed nutrients, with the cheapest source usually being from home-grown fodder or grasses, shrubs, trees and other types of forage grown specifically for livestock feed. These supplies can also include forage by-products from other cash crops, such as sweet corn stover (leaves and stems) from sweet corn grown for sale. Milking cows require concentrates to supplement their forage and some of these can be sourced from by-products from cash crops, such as rice bran or coconut meal.

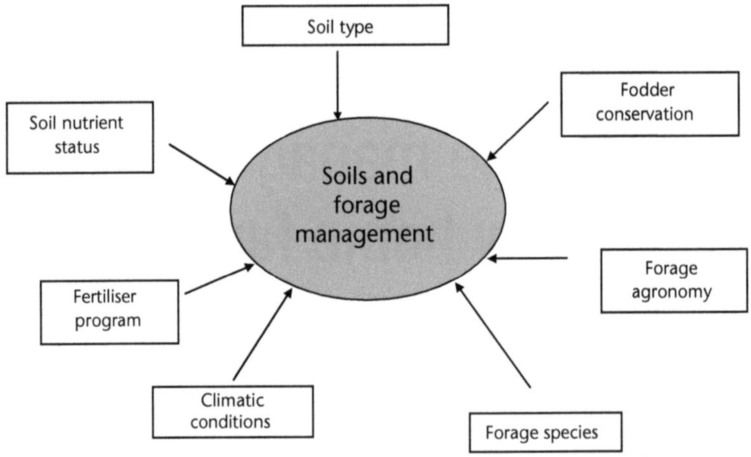

Figure 3.1. The key factors to consider when implementing soils and forage management on the new home farm

Home grown forage is generally the cheapest source of feed nutrients for the dairy herd, because if someone else grows it, they usually require a return for their labour, equipment and transportation, thus charging a higher price than if it was grown on the home farm. In addition, feed nutrients from home grown forages are invariably cheaper than feed nutrients from purchased concentrates and by-products (Moran 2005). Depending on sourcing most of the herd's forage requirements off farm adds yet another uncertainty to the farm program.

Another benefit of home-grown forages is that their management is under the control of the farmer, who can decide how much fertiliser to apply and when to harvest the crop. This should lead to a higher quality fodder: one essential basis of profitable dairy farming.

The basis of economic dairy farming is then producing and using quality forages. To maintain milk composition, milking cows require diets comprising 30–40% forages (on a dry matter or DM basis). Milking cows have very high nutrient requirements and poor-quality forages will just not supply them, because of the physical limitations of rumen capacity. Furthermore, the physical demands of hand harvesting and carrying forages to stalled cows also reduces the likelihood of cows being supplied with sufficient quantities of forages. Because most forages are harvested by hand, tall erect forage species are preferable to prostrate species. Forage quality can be ensured by selecting improved varieties of forages with optimum agronomic practices, such as those described in this chapter.

There are four basic principles of producing quality forages. These are:

1. Select the most appropriate forage species for the region.
2. Prepare the forage production area for sowing.
3. Manage the crop, particularly with adequate fertiliser to optimise growth and quality.
4. Harvest the crop at the best stage of maturity for maximum nutritive value.

This farmer wilts the grass before feeding it to his cows (Indonesia).

3.1 Selection of forage species

The forage must suit the local conditions. Farmers should ask the following key questions when selecting forage species:

- Are they adapted to the climate and soils?
- Do they suit their intended use?
- Do they fit into their particular farming system?

More specific questions to ask are:

- Are there advantages over local varieties?
- Have they been tried and found successful in the region?
- Do they suit local farming systems and ecological conditions?
- What extra inputs are required, such as seed costs, labour and fertilisers?
- Will their extra cost return a profit?
- What are the risks of crop failure?
- Do the seeds come from a reliable source of supply?

No forages grow well everywhere. Some grow well in acidic soils, while others do not. Some grow well in cool areas, while others do not. Forages can survive in areas where they are not adapted, but they will not thrive. It is important to choose forages that are adapted to the local soils and climate.

Important climatic factors affecting forage adaptation are the length of the growing season, temperature, soil fertility, soil pH and drainage.

The key properties of soil that influence plant growth are:

- soil texture: light (sandy, sandy loam), medium (loam, clay loam), heavy (heavy clay)
- soil fertility: low, medium, high. Soil fertility is the ability of the soil to hold and release nutrients for plant growth, rather than the amount of nutrients in the soil at any one time.
- soil pH: strongly acid (<5.0), acidic (5.0 to 6.5), neutral (6.5 to 7.5), alkaline (>7.5)
- drainage: well drained, moderately drained (occasionally waterlogged), poorly drained (frequently waterlogged)
- level of soil salinity: low, medium, high
- level of available Al (aluminium) and Mn (manganese): low, medium, high.

Other important descriptors of forage plants are:

- latitude: tropics, subtropics
- altitude (m above sea level): 0–1000, 1000–2000, 2000–2500, >2500
- rainfall: from 200 to 6000 mm
- family: grass, legume, other (e.g. trees)
- life cycle: annual, perennial
- longevity of forage crop: short term (<4 yr), long term (>4 yr)
- purpose: grazing, cut and carry, conservation, hedgerow or living fence
- defined dry season: <6 months, >6 months
- inundation: 1 week, 1 month, >1 month
- stem habit: erect, prostrate, climbing, stoloniferous (stems growing along the ground), rhizomatous (root nodules that convert atmospheric nitrogen into plant nitrogen)
- shade environment: moderate (30–50% shade), dense (>50% shade)
- frost intensity: light, heavy.

The key forages grown on many SHD farms around Asia include:

- *Pennisetum purpureum* (and hybrids) or Napier (often called Elephant), Mott or King grasses
- *Brachiaria ruziensis* or Ruzi grass
- *Panicum maximum* or Guinea grass
- *Leucaena leucocephala* or Leucaena (legume tree)
- *Gliricidia sepium* or Gliricidia (tree legume).

In recent years, several excellent sources of information have been developed to assist with forage selection for Asian livestock producers. These include:

1. A concise bulletin on selecting forages for livestock feeding in SE Asia by Horne and Stur (1999).
2. The growing and conserving of quality tropical forages for Asian SHD farmers has been covered in detail in Chapters 8 and 9 of *Tropical Dairy Farming* (Moran 2005).

3. A CD presents a process for forage species selection based on their optimum climate, soils, production system and management (Cook *et al.* 2005).
4. A webpage has collated the FAO database of 600 species of tropical forages and legumes (FAO 2005a).
5. A webpage has collated the profiles of 14 Asian countries detailing their climate, livestock and forage resources (FAO 2005b).

3.2 The agronomy of forage crops

The key issues to ensure high yields of quality forage from any forage crop are:

- preparing the soil for sowing, to ensure a good establishment of the crop and minimum weed invasion
- fertilising the crop with sufficient soil nutrients, using inorganic fertilisers to supplement dairy shed effluent and ensuring fertilisers are routinely applied after every (or at least every second) forage harvest
- employing a crop harvesting program to ensure quality forage, as well as adequate forage yields.

3.2.1 Fertilising forage crops

To ensure consistently high yields of quality livestock fodder, forage crops must be 'fed properly' with soil nutrients.

Fertilisers cost money, but they return more through improved yields and quality of forage, hence more milk is produced. Provided other soil nutrients are not limiting plant growth, urea fertiliser can produce an extra 9 kg forage DM/kg urea or 18 kg DM/kg N applied. When harvested and fed to milking cows, this extra forage can yield an additional 9 kg milk/kg urea N (STOAS 1999).

One major limitation of forage production on most small holder dairies throughout Asia is the poor adoption of inorganic fertilisers. Use of cow manure only to fertilise grasses is common practice in most SHD areas, with most farmers not even aware of the economic gains through using inorganic fertilisers. Cow manure supplies organic matter to the forage area, but insufficient N to maximise forage yields and quality. SHD farmers should apply at least 100 kg urea/ha/yr to their forage production area, in addition to the recycled manure (Moran 2005).

Many Western dairy farmers base their fertiliser decisions on the nutrient status of the soils. This is just not possible for most Asian small holder farmers because of the lack of soil testing laboratories and/or the cost of such analyses. However, such advice may be available from local agronomists who service cash crops such as rice or cassava.

A visual and very simple method of assessing the likely response to fertiliser applications is to demonstrate it to farmers through test strips, whereby they apply different fertiliser regimes to small sections of their forage area, such as rows two or three plants wide, to visually assess any response. Erect tropical forages such as Napier grass

A well-managed crop of grass on a Vietnamese small holder dairy farm.

rapidly respond to improved N status by producing dark green coloured foliage plus a more rapid growing, hence taller, plant. By trialling different fertiliser regimes after each harvesting, say nil urea, 50 kg urea/ha, 100 kg urea/ha, urea plus phosphorus fertiliser, and with and without cow manure, farmers can assess whether they should be applying P as well as N. Because they can associate extra, darker green forage with more milk, in most cases, they should respond accordingly by changing their fertiliser management.

3.2.2 Harvesting forage crops

When harvesting forages, farmers tend to place too much emphasis on forage yield rather than forage quality. If forages are too mature and of poor quality, cows might even produce less milk per ha per yr. In the wet season, there is always a compromise between harvesting high yields of low-quality forages (7–8 MJ/kg DM of metabolisable energy or ME) and harvesting lower yields of higher quality forage (9–10 MJ/kg DM of ME). The data presented in Table 3.1 clearly shows that Napier grass must be harvested frequently during the wet season (every 4 weeks) to produce a milking-quality forage.

In a recent review of published data on Napier grass, Muia *et al.* (2000) classified grass quality according to the crude protein requirements for various categories of milking cows. These were for cows at maintenance (5–7%), or producing low (8–10%), medium (11–13%) and high (14–16%) yields of milk. Predicted forage maturity, yield and nutritive value for these various swards are presented in Table 3.2.

High-yielding cows then require Napier grass no more than 42 cm high and harvested every 30 days. Annual pasture DM yields are then likely to be only 60%.

Table 3.1. Quality of Napier grass cut at various stages of regrowth during the wet season

Regrowth (weeks)	Height (cm)	Crude protein (%)	ME (MJ/kg DM)	Total digestible nutrients (%)	Crude fibre (%)
4	50	10.8	9.6	62	28.5
6	75	8.8	8.1	54	32.2
8	135	8.0	7.9	53	32.8
10	150	7.8	7.7	52	33.0
12	150	4.6	7.5	51	31.9

Napier grass harvested for low-yielding cows give annual yields of CP and ME of 105% and 73%, respectively. Therefore, although forage yields will suffer, milk yields will be less adversely affected.

The general recommendations for moderate milk yields (10–15 kg/cow/day) in 450 kg dairy cows consuming 13.5 kg DM/day, is for Napier grass containing 8–13% protein and 7–8 MJ/kg DM of ME. This can be supplied from forage harvested at 42–70 days, when 60–100 cm high, providing a carrying capacity of 4–4.5 cows/ha. For high milk production (<15 kg/cow/day), harvest intervals would have to be increased to 30 days. Milk yields per ha would be high even with low carrying capacity, because of the increased protein yields. However, the increased cost of more frequent harvestings and greater fertiliser applications to maintain soil fertility will reduce the economic benefits of the higher milk yields.

3.2.3 Harvesting tree legumes

With tree forages, regular harvesting of leaves is necessary to maintain favourable leaf-to-branch ratio. For example, *Leucaena* and *Gliricidia* should be harvested every 6–12 weeks. A 12-week harvest interval of *Gliricidia* can yield 9.2 t DM/ha/yr when planted as a block or 1.1 t DM/ha/yr (with forage containing 25% crude protein) when planted as a 40 cm fence around 1 ha of cropping land (Humphries 1999).

Table 3.2. Forage quality and yield of Napier grass cut at various protein contents (Muia *et al.* 2000)

Forage protein content	5–7%	8–10%	11–13%	14–16%
Age (days)	99	63	53	30
Height (cm)	128	95	61	42
DM content (%)	20	17	16	14
NDF content (%)	68	63	61	54
ME content (MJ/kg DM)	7.1	7.7	8.3	8.9
Yield of DM (t/ha/yr)	28.5	21.8	19.7	13.7
Yield of CP (t/ha/yr)	1.7	2.0	2.4	2.1
Yield of ME (000 MJ/ha/yr)	202	168	164	122

It is difficult to recommend specific harvesting regimes for all crops in every situation because the rate of forage growth depends on many factors, such as:

- the forage variety; different varieties have differing optimum harvest intervals
- rainfall and/or irrigation
- soil fertility and additional fertiliser used
- the harvest interval.

3.2.4 Wilting to improve forage intake

The high moisture content of young, freshly harvested forages limits the appetite of milking cows through high loads of water in the rumen. Wilting the forages for up to 24 hr removes intracellular moisture without adversely affecting forage quality, provided the forage does not become heated. Data collected during the wet season in Indonesia (Moran and Mickan 2004) showed that freshly harvested young Napier grass can have DM contents as low as 12% and wilting for 24 hr can increase this to 25%.

Wilting racks can easily be constructed from bamboo and built off the ground to encourage air movement under the freshly harvested forage and so aid water removal from the leaves. Conditioning the thicker parts of the forage, to fracture the epidermal layers, will result in faster rate of water removal from the plant. Forage can then remain on the wilting racks for up to 24 hr.

It is essential to stack the forage loosely on wilting racks in thin layers: no thicker than 10 cm. If the forage heats up during wilting, its nutritive value will decrease, due to plant sugars being used by bacteria and fungi during the heating process. In case of heavy rain, a roof made of tarpaulin or thick plastic should be positioned above the wilting racks, allowing easy removal during periods of sunshine. Rewetting of partially wilted forage requires longer wilting times, which will almost certainly reduce its nutritive value.

To quantify the benefits of wilting, cows in mid-lactation were fed chopped Napier grass plus concentrate (6 kg/cow/day), in Indonesia (Moran and Mickan 2004). Wilting the grass for 8 hr increased intakes of fresh grass from 40 to 50 kg/cow/day and increased milk yields from 14.2 to 15.7 L/cow/day. Assuming the concentrate contained 90% DM, the fresh grass 17% DM and the wilted grass 20% DM, the wilting then increased total DM intakes from 12.2 to 15.4 kg/cow/day, an increase of 3.2 kg/cow/day. Clearly, wilting is a simple and practical method to improve intakes of freshly harvested forages, leading to increases in feed intakes, and hence cow performance.

3.3 Making quality silage

The seasonality of rainfall and soil temperature does not allow for regular growth of forage crops throughout the year. High growth rates during the wet season provide the opportunity to harvest excess forage for conserving and storage to be fed out during the dry season when forage sources are usually in short supply. Conserving forages requires multiple handling: namely cutting, then dehydration and storage, and finally feeding out. Farmers do not like the extra work of handling bulky forages and often do not fully understand the benefits. Consequently, conserving forages on SHD farms is not widely practiced and if tried once or twice, rarely becomes a routine.

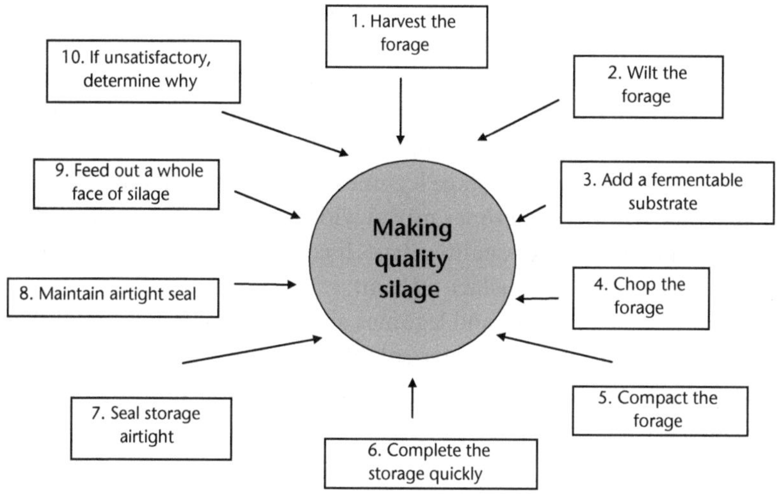

Figure 3.2. The 10 key factors to follow when making quality silage

Excess forages can be conserved as hay or silage. However, ensiling generally produces better quality roughage than hay because less time is required to wilt the feed, when the forage loses nutrients, causing a reduction in feed quality. Hay making requires a longer period of rain-free days, which are often rare in the tropics during the wet season when feed excesses generally occur.

A small holder farmer's silage pit (Indonesia).

The principles of silage making are the same regardless of the size of the operation, the major difference being in the type of storage used (Moran 2005). However, the process or mechanics of silage making (labour, timing and resources) for individual small holders is completely different to those in larger communal farms, where labour and other resources can be shared or amalgamated for efficiencies of size of operation.

All the major forages (grasses, forage legumes, tree legumes and by-products) can be stored as silage. Rice straw is sometimes mixed with very moist forages to reduce effluent losses, but this results in a poorer quality silage. It could be used at the bottom of a silage pit, to absorb the highly polluting silage effluent.

Unfortunately, tropical forages and legumes are not well suited to ensiling due to their inherent low concentrations of water-soluble carbohydrates (that is, sugar, or one of the storage carbohydrates), compared with temperate species. However, rapidly wilting the forage or adding a fermentable substrate, such as molasses before ensiling, will usually result in well-fermented silages.

During the wet season, tropical forage species grow at very fast rates, with forage yields often exceeding animal requirements. If not cut and fed, it will continue to grow, producing very long and fibrous material, which is low in energy and protein.

If this forage was harvested and successfully stored as silage at the same stage as it is cut for producing milk, then it could be fed back during the following dry season. Although the quality of the forage will be slightly lower than its fresh state (10–15% lower in good ensiling conditions), it will still be better quality than many of the forages only available for dry season feeding. Conversely, in some locations, the silage can supplement other good quality, but very slow growing, forages.

Silage can be stored in small plastic bags, 120–200 L plastic and steel drums (plastic lined or bagged), small or large pits dug into hillsides and in stacks above ground. Steel and plastic drums should be stored on their lids to minimise losses when/if air enters the lid. When stored on their base and air enters (as it often will), then decomposition of the silage continues as more air enters. However, when stored upside down, the weight of the silage causes it to drop down inside the drum and minimise air entry.

3.4 The 10 steps to making silage

The success or failure in making quality silage is affected greatly by practices and requires a strict set of guidelines. There are no short cuts! Sections 3.4.1 to 3.4.10 describe the 10 steps to making good quality silage.

3.4.1 Harvest the forage

Forages should be harvested when excess to feed requirements but also when high in quality. In the wet season, tropical forages, such as Napier grass and sorghums, can grow rapidly to heights in excess of 2–3 m. Following this rapid growth, preferably even before it produces more forage than the small holder needs, this excess should be ensiled. This will allow the entire supply of forage to be maintained as high quality into the dry season, rather than it becoming long with a reduced leaf content, and hence low in quality.

It is important that forage harvested for silage should be the same stage of maturity; that is, at its optimum for feeding fresh. For example, Napier grass should be harvested following 30–40 days regrowth in the wet season, at about 75–150 cm in height, for optimum quality and for ease of transporting to livestock in small holdings. At this stage, the Napier grass will have about two to three nodes showing on the stem.

Native roadside forages should be harvested when leafy and contain no prickly species. Tree legumes should be cut while leaves are still green and contain minimal twigs or branches.

3.4.2 Wilt the forage to 30% DM

Tropical species are difficult to ensile due to their high buffering ability; that is, their resistance to changes in pH. To enable them to undergo a more satisfactory fermentation, two techniques are available to small holders: wilting the forage before and/or adding a fermentable substrate at ensiling.

Napier grass will be about 12–15% DM at harvest and should, if possible, be wilted to at least 30% DM. Wilting involves laying the cut forage on racks or against walls to allow the sun's heat to evaporate some moisture from the plants. If rain is likely to fall, the material must then be covered (with plastic or palm leaves) or moved under shelter.

When harvested in the morning, wilting may only require the heat of the afternoon of that day, but when cut later in the day or on cloudy days, it may need wilting till midday of the following day. The layer of material to be wilted should be no thicker than 10 cm, and should be turned over two to three times to encourage wilting. If too thick, the forage will heat and begin to decompose and encourage the wrong types of bacteria to grow. Forage quality and DM will be lost.

3.4.3 Add a fermentable substrate at ensiling

If the fresh forage cannot be wilted, the fermentation of the silage will be improved by mixing the chopped material with 3–5% molasses (on a fresh weight basis) just before ensiling. Although this is a time consuming and messy job, the rewards are well worth it. Adding water to the molasses is not recommended because the forage is already too moist and extra water will just reduce the fermentation quality.

Rather than mixing it in thoroughly, the molasses can be spread as layers in the forage, say every 10–15 cm. Where the molasses is applied, the silage ferments better and is sweeter smelling, but the overall silage quality is still good. Other suitable fermentable substrates include rice bran or formulated concentrates (mixed at 10%) in layers with molasses (5%) poured on top of the rice bran.

3.4.4 Chop the forage into short lengths

The shorter the chop length the better the compaction, so less air is trapped in the forage, resulting in better silage quality. Chop lengths should be from 1 to 3 cm. Mechanised forage choppers will chop quickly to very short lengths. However, small holders can manually achieve similar chop lengths using knives, but this requires high labour inputs and is much slower.

If chop lengths are longer, additional molasses (5–6% on a fresh forage basis) may improve the fermentation. However, it is important that the stems should be chopped to small lengths because they are harder to compact. Leaves can be left at 3–8 cm.

Where the forage has become too long, but is still in a vegetative state (not yet in head), only the leaves and the top end of the stems should be chopped and ensiled, to produce a higher quality silage.

3.4.5 Compact the forage as tightly as possible

Regardless of the system of silage storage, the forage must be compacted as densely as possible – it must be so compact that it is difficult to insert your fingers into the stack. As mentioned above, the shorter the material is chopped, the more dense it can be packed and the less air that will be trapped inside the stack.

If compaction is by human trampling, pockets of air may be trapped inside the stack. The edges of the storage must be well packed. Poles or feet may be used to compress the edges in drums and material must be pushed into corners of plastic bags by hand. Care must be taken not to puncture the plastic bags with fingers, wooden poles, and so on.

Larger stacks of silage in cement boxes or in pits in the ground will require continual trampling while the forage is being delivered. It should be spread evenly and thinly (no more than 5–7 cm thick) over the stack to enable it to pack more densely.

3.4.6 Complete the entire storage quickly

The entire silage storage should be filled and sealed in 1 day, and at a maximum, 2 days. This is easily achieved with bags, drums and small concrete boxes.

In larger stacks, where the forage may require several days to be delivered, the forage from each day should cover that from the previous day to a depth of at least 1 m. The current day's forage then acts as a 'seal' for the previous day. If some of the previous day's forage is not covered sufficiently, it will suffer from aerobic deterioration, causing the stack to heat up, with subsequent losses in both quantity and quality.

Each night until it is filled, the stack should be covered with a sheet of plastic or a thick layer of banana or palm leaves. This will minimise the amount of warm air leaving the stack, which sets up convection currents, thus encouraging more air to enter. This is particularly important with wilted tropical forage, because it is more prone to aerobic deterioration than are temperate forage species.

3.4.7 Seal storage airtight

Silages in well-sealed storages, which prevent the entry of air or water, will maintain their quality for much longer than will silage in poorly sealed storages.

Plastic bags: Forages ensiled inside small bags should be stored inside a second bag because the thin plastic is easily punctured. Furthermore, non-punctured stretched plastic can allow entry of air. To ensure a tight seal, the neck should be twisted and then tied/taped, then doubled over and retied/taped.

Bags must be stored under cover and protected from vermin, rodents, birds, poultry, children, and so on, to avoid punctures. To prevent the plastic breaking down, and to minimise direct heating of the bags, they should also be protected from direct sunlight.

Plastic and steel drums: The tops of the drums should be covered with a sheet of plastic before the lid is placed on top. To ensure an airtight seal once clamped or screwed in place, plastic tape should be placed around the top. The drums should then be stored upside down and preferably under cover or protected from direct sunlight to minimise heating.

Concrete silo or boxes: To reduce losses through aerobic deterioration once opened, it is useful to divide large concrete silos into smaller compartments. This can be done with straw, mud, cement bricks or by using a rectangular timber frame. At least one side of the stack should be sealed with plastic to prevent air entry during feed out. Straw placed above stones in the base of the pit allows moisture to seep to the bottom, but does not allow air entry into the stack. The stones should be approximately 2–4 cm high depending on stack size.

Silage pits or bunkers: Immediately after the filling is complete and the stack well compacted, it should be sealed airtight using plastic, preferably plastic treated with ultraviolet light inhibitor. If such plastic is not available (e.g. if only builders plastic is available), it should be entirely covered with about 10–15 cm of soil to protect it from direct sunlight.

When using ultraviolet-light-treated plastic, it must be weighted with tyres cut in half or sand/dirt filled bags to maintain the plastic close to the silage, thus preventing air entry and movement under the seal. Covering the plastic with soil would be ideal.

3.4.8 Maintain airtight seal

All storage types must be sealed then kept airtight throughout the entire storage. If the plastic is punctured, or stacks start to shrink too much, the cause of air entry into the silage must be determined and repaired as soon as possible.

Effluent flowing out of the storage for longer than 2–4 weeks is indicative that the silage is slowly deteriorating (rotting) due to entry of air. The cause of the air entry should be identified and stopped. If it cannot be stopped, ensure that the same mistake is not made in the future. A mistake is something wrong that happens more than once!

Wilted silage should have little or no effluent unless the stack is poorly sealed. Unwilted silage will produce some effluent, which may leak out of drums and stacks into the soil. Silage effluent should be prevented from entering waterways and drinking water because it causes pollution. It can kill plants or fish if in large numbers.

Only small amounts of silage effluent will leak from well-sealed drums and plastic bags, and may even leak slowly from upturned drums. It is important not to remove drum lids, untie bag tops or hole their bottoms to let moisture out, or to 'see how they are going!' This will allow far too much air to enter, leading to very poorly fermented silages or even just compost.

3.4.9 Feed out a whole face of the storage

As soon as the storage is opened for feeding, air will enter and the silage will begin to deteriorate. Silage in small storages should be fed out completely within 1–3 days of opening. If drums are being fed out over more than 3 days, plastic and weights should be placed over the open face to minimise air entry to the silage.

Unless the forage has been chopped very short (1–3 cm) and well compacted, air enters silage stacks of tropical species very easily. For large silage storages, the whole face of the stack should be removed every day to a depth of at least 20 cm. If the silage is only fed out every 2 days, at least 30–40 cm should be removed every second day. Stack widths should be designed to ensure it takes no longer than 2–3 days to feed out the entire feeding face.

The weighting material on top of the stack should be removed only as required to prevent air moving back into the stack under the top seal. If the silage is warming once opened, it starts to deteriorate and lose yield and quality. If steam is rising from the stack, or if the silage becomes very hot, aerobic deterioration is extreme and the feeding rate must be increased rapidly, unless the problem is due to air entry via other means.

3.4.10 If silage is unsatisfactory, determine the reason

It is important to learn from your mistakes and ensure that the silage is consistently of good quality from year to year.

Silage that has undergone an unsatisfactory fermentation will be unpalatable, and in some cases even poisonous, to animals. Once opened, such silage may be recognised by the following characteristics:

- it has a strong, pungent, very unpleasant smell
- it has a strong ammonia smell
- it contains excess moisture when squeezed or continually oozes from the base
- it is mouldy or slimy
- it has undergone much deterioration (>20% DM loss)
- it is slightly damp and dark brown
- the plastic sheet or lid has not stopped air entry for many days.

Even though animals may not initially eat the spoiled silage, they may do so once the silage has been left in feed troughs for an hour or so, thus allowing some of the pungent smells to escape. However, animal production will suffer as a result. To avoid any possible animal health problems, mouldy silage should never be fed to pregnant and lactating animals.

Silage has a characteristic odour unfamiliar to most livestock. Therefore they may not immediately consume the silage offered without some incentive, such as molasses or fresh forage mixed in with the material. It is very unusual for stock to refuse silage if a slow introductory period has been employed.

Even when they make good-quality silage, farmer adoption rates of silage making have not always been high. As with other feeding technologies such as chemically treated rice straw, farmers do not like to double handle forages. Often, they need to see the benefits (i.e. higher milk yields) of feeding better quality roughages before they accept the higher workloads in silage making rather than purchasing other forages, which are generally of poor quality, for dry season feeding. Another problem is the shortage of excess wet season forage with which to make silage, because they may have high stock numbers on limited areas.

3.5 Storing wet by-products as silage

Fruit, fish waste, vegetables and root crops are increasingly integrated into tropical farming systems and provide a wide range of valuable wet by-products and residues, which are often underused or wasted. The ensiling of such by-products is a simple conservation method and a most effective way to improve animal feed resources through the rational use of such potential feedstuffs to small holder farming.

The major problems usually encountered are the seasonality of supply and their high moisture content. High moisture by-products often have high nutritive value. It is difficult and expensive to dry them, so all too frequently such by-products often become contaminating wastes that quickly go sour, mouldy and lose much of their soluble nutrients as effluent. The advantages of ensiling such material include:

- enabling feeding when such by-products are not being produced
- increasing feed resources and an insurance for high nutrient demands, such as those of milking cows
- reducing demands on home-grown forages
- reducing total feed costs (if low cost)
- improving their palatability
- reducing toxicity to safe levels (in vegetables or cassava leaves)
- destroying harmful bacteria (in poultry litter or fish wastes)
- its use as a major proportion of diets.

The basic principles are the same as those for fresh forages, so attention must be paid to ensuring anaerobic conditions and there should be sufficient acid in the silage to restrict the activities of undesirable bacteria. To achieve successful silage, attention should be given to:

- **moisture content**, which should be at least 50% for ease of compacting to eliminate air. Excessive moisture (>75%), can lead to an undesirable fermentation, producing a sour silage that reduces palatability hence intake. Adding water or using absorbent materials will allow the manipulation of moisture content.
- **length of chopping** – the finer the chopping the better the compaction
- **time to fill the stack** – the quicker the better, and it should be covered each night during filling to reduce invasion of air
- **fermentable energy** – these silages require a stable low pH to minimise biological activity. The final pH depends on the carbohydrate content, which may be sufficient in the material being ensiled or from added sources. For example, protein-rich by-products with low sugar or starch content are difficult to ensile, so should be mixed with energy-rich by-products such as waste bananas, molasses or root crops.

Once opened, every effort must be made to reduce aerobic deterioration. Ensiling in layers separated by plastic sheets can reduce the size of each package of silage. Plastic bags are easy to handle, as well as making excellent mini silos.

Well-made silage can be opened within a month or stored for 6 months or more, provided the cover doesn't break down and allow air to enter the stack.

4

Young stock management

This chapter explains the nutritional needs of heifers in their two phases of development – milk-fed calf and weaned heifer.

The main points in this chapter

- A system of heifer rearing should produce healthy animals that are able to grow to target live weight with minimum input costs.
- It is essential that calves consume at least 4 L of high-quality colostrum in their first 6 hr of life. Calves must be hand-fed colostrum if they cannot suckle.
- Calves should be fed to promote rumen development using high-quality concentrates plus limited roughages.
- Calves can be weaned by 6 weeks of age if they are eating 0.75 kg concentrates/day.
- Calf scours is usually caused by poor feeding management and the first level of treatment should be feeding electrolyte solutions. Antibiotics should not be given until a specific bacterial agent has been identified as a major cause.
- Young calves cannot eat enough forage to sustain good growth rates. Forages alone are not suitable for milk-fed calves or for weaned calves until they reach 200 kg live weight.
- Good heifer growth rates are important for milk production and fertility and to minimise calving difficulties. Growth rate in Friesians after weaning should average 0.6–0.7 kg/day.
- Growth should be monitored regularly (preferably by weighing) to ensure that targets are being met.
- Forages should be high quality (11 MJ/kg DM of metabolisable energy) if it is used as the sole food for heifers less than 12 months of age.

When importing high genetic merit pregnant heifers into small holder systems, the dairy farmer is getting two animals – the pregnant heifer and the unborn calf – for the price of one. If the unborn calf is a heifer bred from a good quality dairy bull, the farmer will have two generations of potentially highly productive milk producers. If the unborn calf

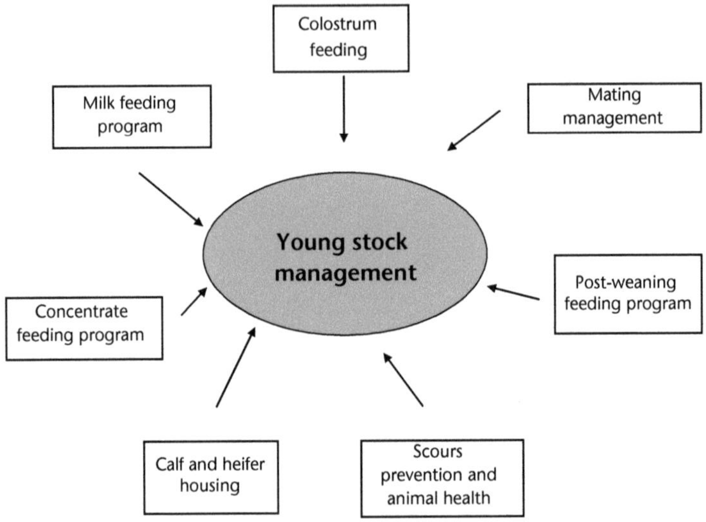

Figure 4.1. The key factors to consider when planning young stock programs

is a bull, it is likely that it will be able to grow at a faster rate than any of the farmer's bull calves born from local stock.

If the imported heifer is purchased as a non-pregnant heifer, then it is important that she be mated to a good-quality dairy bull, either through natural mating or artificial insemination (AI) to ensure her progeny can grow to become another potential high-producing cow in the herd. The emphasis is on the word 'potential' because high genetic merit calves require close attention during rearing to ensure they can achieve close to their genetic potential once they calve and join the milking herd.

Unfortunately, far too many dairy farmers in SHD herds restrict the potential of their newborn calves (Moran 2012). This results in the all too common observation that the second generation of imported stock are invariably less productive than their dams. Because this is entirely the result of poor calf rearing practices, this chapter is essential reading for all dairy farmers because it provides the groundwork to improve the ultimate performance of their heifer calves.

Well-grown dairy heifers are a good investment in the milking herd. To ensure they grow to become productive and efficient dairy cows, their management must be carefully planned and begins the day they are born.

A well-managed heifer rearing system aims for:

- good animal performance, with minimal disease and mortality
- optimum growth rates to achieve target live weights
- minimum costs of inputs, such as feed (milk, concentrates and forages), animal health needs (veterinary fees and drugs) and other operating costs (milk-feeding equipment) to achieve well-reared heifers
- minimum labour requirements
- maximum utilisation of existing facilities such as sheds for rearing and quality forages for feeding.

A well-managed calf-rearing shed (Sri Lanka).

There is no single best way to milk rear calves. All sorts of combinations of feeding, housing and husbandry can be successful in the right hands and on the right farm. Successful calf rearing is a specialist job requiring suitable facilities. It also requires a genuine concern for the welfare of young calves and quick responses to early symptoms of disease. If farmers are unable to commit the time and resources to rearing their own replacement heifers, they should seriously consider paying someone who is better placed to do a good job. The establishment of cow colonies in many Asian dairy industries, and the sharing of resources of farmers supplying a single dairy cooperative, provide the opportunity for a specialist calf and heifer rearer to milk rear and grow out the replacement stock for many farmers.

The first 3 months are the most expensive period in the life of any dairy cow. During that time, mortality rates are high, up to 20% in many cases. Calves need protection from the extremes of sun, wind and rain, no matter what the rearing system. Disease prevention and treatment can be costly during early life (Moran 2002).

Deaths of young stock should not exceed 4–6%, while the incidence of calf sickness should not exceed 10%.

4.1 Rearing the milk-fed calf

With their undeveloped digestive tract, calves require the highest quality and the most easily digestible source of nutrients: namely, whole milk or milk replacer. Unfortunately, these are also the most expensive feeds. The most effective way of minimising the high feed costs of calf rearing is through early weaning and reduced milk feeding.

The essence of good calf rearing depends on two major nutritional factors. Firstly, an adequate intake of high-quality colostrum within the first day of life and, secondly, feeding management to encourage early rumen development.

4.1.1 Colostrum feeding

Newborn calves are very susceptible to disease. Before they develop their own immunity, they are entirely dependent on the antibodies contained in their mother's milk. It is therefore vital that they receive adequate quantities of antibody-rich colostrum from their mothers or from other freshly calved cows.

Calves should have access to 4 L of colostrum within the first 6 hr of life. They will not need any additional milk for the next 12–24 hr. Any calf that is suspected of not having suckled in the first 3–6 hr should be hand-fed the colostrum. With sick or weak calves, colostrum may have to be administered by stomach tube. It is not difficult to stomach-tube young calves.

Frozen colostrum (which can be stored for 18 months) can be thawed out and used, or colostrum from a mature cow within the herd can be fed. Fresh colostrum can be refrigerated for 7–10 days. Ideally, colostrum should be tested for antibody concentration to ensure it is of sufficient quality to store.

The level of immunity passed on by the cow increases with her age, because older animals have been exposed to a greater range of infectious organisms to which they have developed antibodies.

Replacement heifers born to first calving cows may require additional stored colostrum from older cows to ensure they develop good disease immunity. Vaccinating cows prior to calving for *E. coli*, *Clostridia* and *Salmonella* can enhance the immune properties of colostrum.

The longer calves spend with their dams, the greater their chances of contracting disease. Therefore newborn calves should be separated from their dams as quickly as is practically possible.

4.1.2 Early rumen development

The rumen is non-functional in newborn calves; hence, all digestion must take place in the abomasum (or true stomach) and the small intestine. The weaned calf needs a fully functional rumen in order to be well adapted to a forage-based diet. Before weaning, it is important to promote rumen development, so as to avoid growth checks when calves are weaned.

Rumen development occurs through the digestion or fermentation of feeds (roughages and concentrates) by the rumen microbes. Calves should be encouraged to eat solid feeds at an early age, mainly through limiting their access to milk to 4 kg/day. From the first week, roughage such as clean straw should be offered in combination with high-quality concentrates specially formulated for rearing calves.

Fresh forages are not good sources of roughage for milk-fed calves. Such forages contain too little fibre, and their very high water content prevents high intakes of feed energy in each mouthful. This limits the feed energy available for rapidly growing animals. Until their rumen capacity is larger, young calves just cannot eat enough fresh forage to sustain high growth rates.

4.2 A successful early weaning program

Dairy farmers generally want to feed their calves on the best quality feeds to give the calves a good start to life. On the whole though, most farmers feed too much milk for too long.

By continuing to feed milk longer than is necessary, farmers often feed 400–500 kg of milk (or its equivalent in milk replacer solution) to each calf. They need only feed 200 kg of milk or less. Furthermore, much of this milk can be colostrum from calving cows, making the milk feeding costs very low.

If calves are strong, healthy and kept warm and dry, they can be successfully reared on a once daily feeding with 4 L of whole milk, or its equivalent in milk replacer. All calves should be offered a specially formulated calf meal from 1 week of age. Milking cow concentrate formulations do not contain sufficient protein to meet the needs of young calves.

Calves should have limited access to fresh forages. The key to this rearing system is giving the calves continuous access to clean straw as a source of roughage. Note this is clean straw, not good quality pasture hay or lucerne hay.

All calves must be given the opportunity to nibble on the straw, even though they will eat very little of it. Straw will encourage rumen development rather than provide nutrients.

If better quality hay is fed in place of straw, calves will eat more roughage but at the expense of concentrate consumption. If good quality hay is fed, it should be limited to 100–200 g/calf/day. Clean drinking water must be available at all times.

Feeding milk only once each day helps the calves to develop an appetite for the concentrates. It is the concentrate, rather than the milk, that should provide the bulk of nutrients to keep the calf growing. Calves can be weaned off milk once they are consuming 0.75 kg/day of concentrates for 2–3 consecutive days. This usually occurs by about 6–8 weeks of age.

Provided calves are eating 0.75 kg/day of concentrates, milk feeding does not have to be reduced gradually. Calves should continue to be housed indoors during weaning.

This system rears the rumen rather than the calf. Systems that involve feeding more milk do not encourage early rumen development and hence calves must be older before they can continue to grow without milk.

It is important that each calf drinks its allocation of milk. Lower milk intakes will limit calf performance because of the inability of the young animal to compensate by eating more concentrates. As well as reducing growth rates, underfed calves may be more susceptible to diseases and other stresses during life. Higher milk intakes will discourage concentrate consumption.

Ideally, calves should be housed individually or in small groups. They should also be individually bucket-fed. There is no advantage in milk feeding using teats rather than buckets; it only creates extra work in keeping the buckets clean.

4.2.1 Weaning age

The age when milk is no longer fed should depend on the quality of feeds available. For example, Ibrahim (1988) suggested calves should be weaned at:

- 2 months, when quantity and quality of roughage and concentrates are good
- 4 months, when quantity and quality of roughage and concentrates are average
- 6 months, when quantity and quality of roughage and concentrates are poor
- 8 months, when suckling and cows are dried off.

Provided they are not adversely affected by disease or very poor feeding management, the growth rates of calves up to weaning is far less of an issue than their performance post-weaning. Prior to weaning, calves are fed milk, which contains all the essential nutrients for healthy growth, whereas after weaning the quality and amount of dry feeds have the biggest influence on calf performance. Invariably farmers save their best quality feeds for their milking cows, meaning that young stock suffer from poor feeding management. Weaning programs developed for young stock in temperate regions, where feed quality is usually better, therefore require close scrutiny to ensure they will still allow weaned stock to achieve their target growth rates (see Table 4.2).

4.2.2 Concentrate quality

Milk-fed and weaned calves require concentrates containing higher protein levels (18–20%) than do milking cows (16%). Low protein concentrates will not promote the same rate of rumen and body development in milk-fed calves. Consequently 2 months may be too young for weaning such calves.

Throughout Asia, the majority of formulated supplements for milking cows are formulated to 16% protein, even though on closer investigation (Moran 2005), they are frequently below this content. For convenience, many SHD farmers also feed these concentrates to their young stock. Such formulations are far from ideal because of the higher protein requirements of young stock. Depending on the quality of the basal roughage fed post-weaning, 18% may even be insufficient.

Very rarely can SHD farmers purchase higher protein formulated concentrates and in many cases farmers are not even aware of the benefits for their young stock in supplementing available milking concentrates with additional protein supplements.

High protein concentrates may be available, but at great expense, because they have been formulated for pig and poultry, incorporating high-quality protein ingredients. It would be ideal if a few large-scale feed mills, either owned by dairy cooperatives or agribusiness, could formulate calf and heifer mixes with higher protein contents, using better quality energy sources and additional minerals and vitamins for optimal growth of young stock. Compared with the higher demand of concentrates specially formulated for milking cows, the formulation of smaller batches of calf/heifer mixes would not be cost effective for small dairy cooperatives.

4.2.3 The high cost of milk rearing

Most Asian small holder farmers feed milk (or milk replacer) to their calves for 8–12 weeks. This is the 'accepted method' because the high nutrient intakes should ensure a

Table 4.1. The cost of dietary energy when rearing dairy calves in Sri Lanka

	Cost (Rs/kg)	Cost for DM (Rs/kg)	Cost for energy (Rs/MJ)
Whole milk			
Expensive	35	292	13.3
Cheap	20	167	7.6
Calf concentrate			
Expensive	35	39	3.1
Cheap	25	28	2.2

good growth rate, provided the calves remain healthy, because specially formulated calf concentrates are not readily available.

Feeding high levels of milk can cause several problems. Firstly, it is invariably more expensive to provide the same intake of feed nutrients from milk or milk replacer than from formulated calf concentrates.

Table 4.1 is based on whole milk containing 12.0% DM and 22.0 MJ/kg DM of energy and calf concentrates containing 90% DM and 12.5 MJ/kg DM of energy. The calculations are for Sri Lanka small holder farmers (using Sri Lankan rupees or Rs as the currency), but they provide a guide to farmers in any country.

For farmers receiving 35 Rs/kg for their milk, they are spending four times more per unit of energy by continuing milk feeding beyond the 6–8 weeks required for normal rumen development in a well-reared calf.

The second problem with high milk feeding is the increased likelihood of disease problems. Once milk is removed from their diet through weaning, calves are more resistant to scours. Unless a strict cleaning and sterilising routine is enforced in the calf shed, flies and other disease carrying agents will thrive on residual milk left in buckets, on floors and in other equipment used with milk feeding.

4.3 Scours in milk-fed calves

Animal health will be discussed in a later chapter, but any discussion of calf rearing must include calf scours (or neonatal diarrhoea). This is a major problem throughout Asian small holdings because many veterinarians just do not understand its causes, hence its prevention and treatment (Moran 2012). Antibiotics should be given as the last resort of treating scours, rather than administered routinely to all scouring calves.

Scours is the single most important cause of death in milk-fed calves. Even when calves survive, the increased labour requirements for their care, together with veterinary and drug bills, make scours a costly problem with calf rearers. Furthermore, their potential productivity is likely to be decreased by setbacks in early life.

The causes of scours, particularly in calves under 3 weeks of age, are complex. There is not usually a single cause, with contributing factors being calf management, diet, the environment and the presence of pathogens. Prevention and early treatment with fluid replacement (electrolyte therapy) can avoid the unnecessary use of antibiotics.

4.3.1 What is scours?

Calf scours is easily recognised. It is an increase in the frequency and quantity of faeces, which has a higher than normal water content. In some cases, blood and mucus may also be present. Whatever the cause of scours, farmers will see some or all of the following:

- bright yellow or white faeces
- depressed calves that are reluctant to feed or suck
- calves with sunken eyes and/or a temperature
- skin remaining peaked or tented when lifted, indicating dehydration
- weight loss and weakness
- if severe cases, calves will collapse, become comatose and die.

Scours can be classified into two types, nutritional and infectious. Nutritional scours is usually caused by stress to the calf due to a breakdown in management routines. Nutritional scours often progresses to an infectious scour, which is caused by a high population of pathogens.

4.3.2 Causes of nutritional scours

Scours can be traced back to two major causes: poor colostrum feeding management and stress.

One of the first effects of stress in calves is a reduction of acid secretion into the abomasum. This reduces both the ability of the clot to form, and digestion of milk protein.

Stress can result from a wide variety of causes. It could be due to inappropriate milk feeding programs (e.g. over-feeding or irregular feeding), sudden changes in the concentration of milk replacers, incorrect milk temperatures or poor-quality milk powder. Calves reared on milk replacers are more prone to scours than those reared on whole milk.

Environmental stress is also a common cause of scours, such as sudden changes in the weather (for calves run outdoors) or cold, damp, draughty or humid conditions inside calf sheds. Overcrowding in calf sheds can result in outbreaks of scours. Even changes in staff can lead to scours through different handling of calves or changes in the standards of hygiene. Combinations of any of the above stresses will greatly increase the likelihood of scours.

The duration of scours is largely under the control of the calf rearer. During their second week of life, calves are particularly susceptible. By careful observation, it is possible for experienced calf rearers to anticipate the onset of scours on the day before it occurs, after which milk feeding can be reduced, with the result that the calf recovers quickly.

The following signs of impending scours should be looked for:

- dry muzzle, thick mucus appearing from the nostrils
- very firm faeces
- refusal of milk
- a tendency to lie down
- a high body temperature (over 39.3°C).

Scouring calves can lose up to 5 L of fluid each day including minerals salts essential for normal body function. With most infectious scours, it is the dehydration and acidosis, not the infection, that kills the calf. In fact, most calves can naturally overcome infectious scours if their dehydration is quickly treated. For viral infections, this is the only treatment option.

The loss of electrolytes reduces the ability of body tissues to retain water, thus aggravating the dehydration. This highlights the importance of quick recognition and treatment of scouring calves with electrolyte therapy to replace their lost body fluids.

4.3.3 Treating scours

To treat scours properly, its cause must be determined and eliminated. For example, if it results from poor shed design, a calf rearer may not be able to immediately rectify that problem, although temporary measures such as reducing draughts or decreasing calf numbers should be possible.

The top priority for treating scouring calves is to provide them with sufficient liquid and electrolytes to replace those lost in the faeces. The next priority is to supply additional sources of readily digestible energy, such as glucose (dextrose), but not sucrose. Finally, and only after diagnosis, should drugs be considered for treating the actual pathogens causing infectious scours.

There are a variety of electrolyte fluid replacers on the market, some of which contain a supply of energy, while others may contain chemicals to reduce the rate of passage of feed through the gut or to reduce the rate of loss of body fluids from the tissues into the intestines. Formulations incorporating antibiotics are only effective against scours caused by bacteria and/or protozoa, whereas many of the infectious scours are caused by viruses. Traditional thinking on treating scours was that milk feeding should cease, or at least be reduced. However, milk contains the best source of energy for calves, while some electrolyte treatments contain insufficient energy to maintain a sick calf. It is dangerous to withdraw milk for more than 24–48 hr because the intestinal wall will degenerate and lose its capacity to secrete enzymes that digest lactose. Many authorities now recommend withholding milk for no more than 24 hr, or even not at all.

Scouring calves should be isolated in a clean, dry and warm pen. Frequent small feeds of electrolytes or milk are better than fewer larger ones. Ideally, the daily volume of fluids required should be determined, based on the size of the calf, the degree of dehydration that needs to be corrected and the rate of continuing fluid loss via the faeces.

A decision must then be made as to how to administer the electrolytes. If the calf is able to suck, it can be fed by bucket or teat. If it is unable to drink, it should be fed by stomach tube. If it has collapsed and is in shock, it will require subcutaneous or intravenous therapy.

Diarrhoea powders containing kaolin, pectin or chalk can reduce the severity of scouring. Other home-made treatments, used in the past to slow down the passage of feed through the gut (such as charcoal tablets, cornflour or even sawdust), can make the condition worse and should be avoided. Antibacterial compounds and antibiotics (e.g. calf scour tablets, drenches or injections) should be used only on recommendation of veterinarians and restricted to cases where *E. coli*, *Salmonella* or other bacteria have been diagnosed or are suspected. Prolonged use of antibiotics can lead to nutritional scours by killing off *Lactobacilli* (the milk-digesting bacteria) in the abomasum and preventing development of rumen microflora.

Up to 90% of treated calves should be back to normal after 2 days of fluid therapy, while the remainder may require veterinary examination. This often includes faecal sampling to identify any infectious agent causing the scours. Recording the ages of scouring calves will assist veterinarians in diagnosing the causative pathogen.

When reintroducing milk after fluid therapy, it should be offered full strength. Milk should never be diluted with electrolyte solutions, because this can lead to poor milk clotting. If giving both electrolytes and milk, electrolytes should be given at least 30 min before a milk feed.

4.3.4 Preventing scours

To ensure healthy and disease-resistant calves, the importance of good colostrum feeding management cannot be overemphasised. Up to 40% of calves do not absorb sufficient antibodies within the first 12–24 hr of life because of inadequate attention to their colostrum feeding. Such calves are more likely to succumb to infectious scours.

Feeding high-quality colostrum for the first few days of life is beneficial. Even if some of these antibodies are not absorbed into the blood, they can still provide local protection in the intestines against infectious scours.

Prevention of scours centres around good hygiene and minimising stress. Measures that can be taken include:

- ensuring calves are protected from extremes of climate
- carefully planning shed designs to avoid overcrowding
- minimising stresses associated with routine management practices, such as disbudding and castration
- maintaining strict hygiene by cleaning and sterilising feeding utensils and facilities during milk rearing
- developing a routine milk feeding program, with as few people involved as possible
- weaning early to minimise the period of milk feeding
- quickly responding to early symptoms of scours, isolate sick calves and address the cause
- minimising the use of antibiotics and then only under veterinary supervision
- keeping records of treatment of sick calves to assist in veterinary diagnoses and for withholding periods if the calf is subsequently culled.

4.4 Management of weaned replacement heifers

All too often, farmers rear their heifer calves carefully until weaning but neglect them thereafter. Calves that are poorly managed after weaning are disadvantaged for their entire life. Even if they are well fed after mating, their ultimate mature size is restricted and if they do put on extra weight, it tends to be fat. Most of the growth in skeletal size occurs before, not after, puberty.

Weaned heifers do, however, require less attention than milk-fed calves and milking cows. Dairy heifers need to be well fed between weaning and first calving. If growth rates are not maintained, heifers will not reach their target live weights for mating and first calving.

Undersized heifers have more calving difficulties, produce less milk and have greater difficulty getting back into calf during their first lactation. When lactating, they compete poorly with older, bigger cows for feed. Because they are still growing, they use feed for growth rather than for producing milk. Many studies have demonstrated the benefits of well-grown heifers in terms of fertility, milk production and longevity.

4.4.1 Fertility

The onset of puberty, and commencement of cycling, is related to live weight more than to age. A delay in puberty means later conception. All heifers should achieve their target weight before joining, because lighter heifers have lower conception rates. Calving problems depend more on heifer live weights at mating, than on live weights or body condition at calving. Frame size is determined early, so there is doubtful merit in the practice of feeding older heifers to make up for poor growth earlier in life.

Studies showed that Friesian heifers mated below 260 kg had 34% conception to first insemination compared with 58% for heifers mated weighing 300 kg or more. Of the smaller heifers, 24% had difficult calvings. This declined to 8% in heifers mated at 260–280 kg and was lowest in 340–360 kg heifers. Heifers underweight at mating required considerable assistance if in difficulty during calving.

Contract heifer rearing in Thailand.

4.4.2 Milk production

Increasing calving live weights for Friesians from 360 kg to 460 kg increased milk production during the first lactation by 400 kg (Freeman 1993). This production benefit extended into both the second lactation, with an extra 830 kg/extra 100 kg live weight, and the third lactation, with an extra 840 kg/extra 100 kg live weight. Heifers calving 100 kg heavier can increase their peak production by 5 kg/day during the first lactation.

However, there is little point in rearing well-grown heifers then underfeeding them during their first lactation. Bigger heifers have higher maintenance requirements, which must be met before additional nutrients produce milk. Therefore good heifer-rearing systems should be considered only after feeding systems for milking cows have been developed. Many Asian small holder farmers do not feed their cows well enough to justify producing bigger heifers.

4.4.3 Heifer wastage

Poorly grown heifers do not last long in the milking herd. They are more likely to be culled for poor milk yield or poor fertility during their first lactation.

Total herd costs can be greatly increased by this high rate of wastage. Producers should aim to lose (through deaths or culling) no more than 20% of their replacement heifers between weaning and their second lactation.

4.5 Targets for replacement heifers

4.5.1 Live weight

Traditional target weights are too low to ensure first lactation heifers achieve their productivity potential, particularly on farms where milking cows are well fed. Table 4.2 summarises recommended live weights for Friesian and Jersey heifers at various ages. Targets for Zebu or local breed heifers would be similar to those for Jerseys.

Puberty occurs in dairy heifers at 35–45% of mature weight, while conception can occur at 45–50% of mature weight. A dairy cow will attain her mature live weight in about the fourth lactation and the objective of rearing heifers is to produce an animal 80–85% of mature live weight by first calving. Wither height sticks or chest girth tapes are an alternative to scales, but they are not as accurate and tend to overestimate live weights.

4.5.2 Wither height

Wither height (or height at the shoulder) is a good measure of bone growth and frame size in heifers. Frame size can influence ease of calving and appetite of milking cows.

Table 4.2. Target weights for Friesian and Jersey heifers at different ages (Moran 2002)

Age (months)	Friesian live weight (kg)	Jersey live weight (kg)
2–3 (weaning)	90–110	65–85
12	250–270	200–230
15 (mating)	300–350	250–275
24 (pre-calving)	500–520	380–410

Farmers should aim for wither heights in Friesians of 123–125 cm at 15 months of age and 133–135 cm at 24 months. Corresponding wither heights in Jerseys (and Zebus) are 110–112 cm at 15 months and 120–122 cm at 24 months.

4.5.3 Age of teeth eruption

It is easy to estimate the approximate age of a heifer by inspecting the state of her teeth. A calf may be born without teeth, with the temporary cheek teeth erupting within a few days and the temporary incisor teeth within 2 weeks.

The age at which the pairs of permanent incisor teeth erupt is as follows:

- First incisor teeth: 18–24 months
- Second incisor teeth: 24–30 months
- Third incisor teeth: 36 months
- Fourth incisor teeth: 40–48 months

The permanent cheek teeth erupt between 6 and 36 months, but are harder to identify than the incisor teeth. The age of eruption of permanent incisor teeth can vary with feeding regime.

This is a very useful guide when objectively assessing the feeding management of young stock because poorly fed heifers may look healthy and relatively well grown, but if their first (or even second) incisor teeth have erupted they are likely to be much older than at first glance.

4.5.4 Energy and protein requirements for heifers

Table 4.3 shows the energy requirements (for maintenance and growth) of heifers growing at different rates at various live weights. The growth rates for 500 kg heifers assume a contribution of 0.4 kg/day from the growing foetus.

See Chapter 5 for a description of ME (metabolisable energy). Growing heifers require a constant source of protein for optimum bone and muscle growth. Table 4.3 also lists crude protein requirements at different live weights.

4.5.5 Feeding heifers to achieve target live weights

Recommendations for grazing and feeding systems will differ between regions. Rather than depend on recipes, producers should regularly weigh their young stock, then vary feeding strategies according to their growth rates. Growth should average 0.6–0.7 kg/day,

Table 4.3. Energy and protein requirements for growing heifers (NRC 1989)

Live weight (kg)	Energy requirement MJ/day of ME				Crude protein (%)
	Growth rate (kg/day)				
	0.4	0.5	0.6	0.7	
100	28	30	33	36	17
200	42	45	48	51	16
300	50	60	64	68	15
400	75	79	84	89	13
500	91	95	99	103	13

although that can vary between 0.5 and 1.0 kg/day, depending on available pasture and the supply and cost of suitable supplements.

As fresh forage is the cheapest feed, it should constitute the bulk of the diet, with hay, silage or concentrates used to overcome forage shortages. Fresh forages or conserved hay or silage must be of sufficient quality (at least 10 MJ of ME) to satisfy the requirements for growth and maintenance.

Until calves reach 200 kg in weight, they are not able to maintain the growth rates needed to reach target weights on diets of either average quality forages or even top quality silage. Their capacity is limited and they simply cannot eat enough DM from the forages to meet their nutrient requirements for rapid growth. Forages must be good quality (at least 11 MJ of ME) if used as the sole feed for heifers less than 12 months of age.

Forage quality and allocation should allow for continuous growth throughout the first 2 yr. Uniform growth is not necessary and may be impracticable with seasonality of quality forage supplies. Yearling heifers have some ability for compensatory gain following periods of mild undernutrition, as long as they have not been grossly underfed. However, heifers should not be allowed to lose weight or to grow very slowly for long periods of time (i.e. no more than 1 month).

Ideally, growing heifers should be continuously fed concentrates to supplement the fresh forage, the quantity offered depending on target growth rates and the nutrients provided from the forage.

5

The principles of dairy nutrition

This chapter explains the key principles of dairy nutrition to better understand the practices of feeding management.

The main points in this chapter

- Feeds contain the nutrients for animal survival and production: the most important ones being water, energy, protein and fibre.
- There are several ways to describe the energy concentration of feeds, with metabolisable energy being the preferred method.
- Feed protein and fibre can also be described in various ways.
- Feeds can be categorised into various types depending on their nutrient composition. One good classification system is based on their energy and protein contents.
- It is possible to calculate the nutrient requirements of milking cows to produce target levels of milk.
- Understanding the lactation cycle, from one calving to the next, is important when planning and managing the herd's annual feeding program.

The essence of good farm management is to supply sufficient farm inputs for a desired level of production of farm outputs. For SHD farmers this means setting realistic milk production targets, each day, each month or each year and determining the necessary farm requirements to achieve these targets. The major requirements for sustainable milk production are healthy, productive and fertile dairy cows and their required feed nutrients supplied in forages and concentrates. The forages are either grown on farm or purchased. The concentrates are usually purchased, either already formulated or as their ingredients, the latter usually as agro-industrial by-products.

Other chapters in this manual discuss the important aspects of good dairy herd management, while the following two chapters discuss their feeding management. This chapter explains the principles of dairy nutrition, while the following chapter discusses the feeding management of the milking herd.

Feeding management includes both the supply of feeds and their presentation to the milking cow. The chapters on soil and forage management and on silage making deal

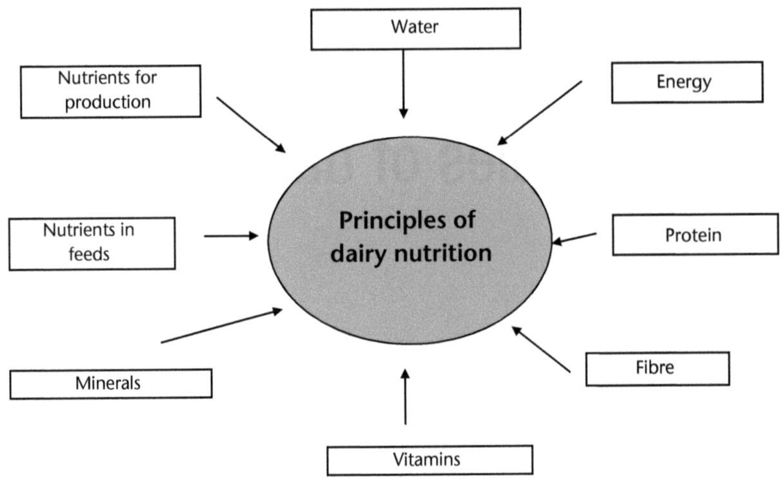

Figure 5.1. The key elements of dairy nutrition

with on-farm feed supplies. This chapter discusses the processes involved in calculating how much feed is required to achieve milk production targets.

Dairy cows are herbivores and have digestive systems well adapted to forage-based diets. Belonging to a group of mammals called ruminants, they have a multi-purpose digestive system. As well as a stomach that breaks feeds down into their basic nutrients, they have a rumen, which is like a large 'fermentation vat' where micro-organisms pre-digest these feeds. The rumen of a mature dairy cow is up to 200 L in volume where the feeds can spend 12 to 24 hr before moving onto the 'true' stomach.

5.1 Nutrients supplied by feeds

When cows consume their smorgasbord of feeds, the major nutrients they extract from them are water, energy, protein, fibre, vitamins and minerals. The four nutrients that have the major influence on cow performance are water, energy, protein and fibre.

1. **Water.** The body of a dairy cow is composed of 70–75% water. Milk is about 87% water. Water is not a feed as such because it does not provide specific nutrients. However, it is essential in body processes and to regulate body temperature. Water is involved in digestion, nutrient transfer, metabolism and waste removal. Water has structural and functional roles in all cells and all body fluids. An abundant, continuous, and clean source of drinking water is vital for dairy cows.
2. **Energy.** Dairy cows use energy to function (to walk, graze, breathe, grow and put on body condition, lactate and maintain a pregnancy). Energy is the key requirement of dairy cows for milk production. It determines milk yield and milk composition.
3. **Protein.** Protein is the material that builds and repairs the body's enzymes and hormones, and is a constituent of all tissues (muscle, skin, organs and foetus). Protein is needed for the body's basic metabolic processes, growth and pregnancy. Protein is also vital for milk production. Proteins are made up of various nitrogen-containing

amino acid molecules. Amino acids are the building blocks for the production of protein for milk, tissue growth and the development of the foetus during pregnancy.

4. **Fibre.** For efficient digestion, the rumen contents must be coarse with an open structure and this is best met by the fibre in the diet. Fibre contains most of the indigestible parts of the diet. It ensures that the cow chews its cud (ruminates) enough and therefore salivates. Saliva buffers the rumen against sudden changes in acidity.
5. **Vitamins.** These are organic compounds that all animals require in very small amounts. At least 15 vitamins are essential for animals. Vitamins are needed for many metabolic processes in the body, such as the production of enzymes, bone formation, milk production, reproduction and disease resistance.
6. **Minerals.** These inorganic elements are needed for teeth and bone formation, enzyme, nerve, cartilage and muscle function or formation, milk production, blood coagulation and efficient utilisation of energy and protein.

5.1.1 Describing feed energy

There are various ways to describe the energy content in feeds and these are briefly discussed below.

1. **Digestibility.** Measured as a percentage, digestibility relates to the portion of food that is not excreted in the faeces and so is available for use by the cow. Digestibility is not a direct measure of energy, but it does indicate overall feed quality. Because cows are able to digest and use more of it, the greater the digestibility, the greater the benefit of that food to the cow.
2. **Metabolisable energy (ME).** This describes the energy in a feed that cows can actually use for their metabolic activities; that is, maintenance, activity, pregnancy, milk production and gain in body condition. The ME content of a feed can be calculated directly from its digestibility. The ME content of a feed (also called its energy density) is measured as megajoules of metabolisable energy per kg of dry matter (MJ ME/kg DM). Intake of ME is expressed in MJ/day. The higher the energy content of a feed, the more energy is available to the animal. If a feed contains 10 MJ/kg DM, then each kg of dry matter of that feed contains 10 MJ of metabolisable energy available for use by the cow. A feed containing 12 MJ/kg DM then has higher energy content than a feed containing 10 MJ/kg DM.
3. **Total digestible nutrients (TDN).** This is an alternative method to describe feed energy. This is a very old energy system, but still used in the US and some Asian countries. TDN is a less accurate measurement of energy than ME, because it does not take into account energy losses via methane (from rumen digestion) and urine. TDN content is expressed as a percentage, with TDN intake expressed in kg/day. TDN and ME are interchangeable.

References to the energy density of feeds are given in either of these two measures, as ME (as MJ/kg DM) and TDN (as %). References to cows' energy requirements or intakes are also given in these two measures, either as ME (MJ/day) or TDN (kg/day). This manual will use the ME system because it is more frequently used by dairy nutritionists.

5.1.2 Describing feed protein

There are various ways to describe dietary protein and these are briefly discussed below.

1. **Crude protein (CP).** Dietary protein is commonly termed 'crude protein'. This can be misleading, because crude protein percentage (CP%) is not measured directly, but is calculated from the amount of nitrogen (N%) in a feed using the formula $CP = N \times 6.25$.
2. **Non-protein nitrogen (NPN).** NPN is not actually protein, it is simply nitrogen. Rumen microbes use energy to convert NPN to microbial protein for use in the body. In forage-fed cows, the rumen microbes use NPN with only 80% efficiency (compared with true protein), which reduces its overall value as a protein source. Urea is a source of NPN.
3. **Rumen degradable protein (RDP).** This is any protein in the diet that is broken down (digested) and used by the microbes in the rumen. If enough energy is available in the rumen, some of this RDP will be used to produce microbial protein.
4. **Undegradable dietary protein (UDP).** This is any protein in the diet that is not digested in the rumen. It is digested 'as eaten', further along the gut. That's why UDP is sometimes called 'bypass protein'. The proportion of the protein that is digested in the rumen is called its degradability.

5.1.3 Describing feed fibre

There are various ways to describe dietary fibre and these are briefly discussed below.

1. **Neutral detergent fibre (NDF).** This is a measure of all the fibre (the digestible and indigestible parts) and indicates how bulky the feed is. Some of it is digested, and some is excreted. A high NDF might mean lower intake because of the bulk, while lower NDF values lead to higher feed intakes.
2. **Acid detergent fibre (ADF).** This is the poorly digested and indigestible parts of the fibre; that is, the cellulose and lignin. If the ADF is low, the feed must be very digestible (i.e. high quality).
3. **Crude fibre (CF).** Although sometimes used to indicate fibre content, CF is now considered an unacceptable measure because it does not take into account the digestible fibre, which is nutritionally useful to the animal, both as a source of energy in the diet and as a substrate for some of the rumen bacteria. However, it is commonly analysed because it is required in the calculation of TDN.

5.2 Nutrients contained in feeds

Most dairy cattle nutritional reference books contain tables of the nutritive value of feeds and supplements. The following two tables provide a summary of some of the common forages (Table 5.1) and supplements (Table 5.2) fed to dairy cows in Asia (Moran 2005). The values are in ranges to take into account differences in the agronomic growing conditions and stage of maturity in forages and the degree of processing of by-products. Table 5.3 classifies feeds on the basis of their average CP and ME contents.

Table 5.1. Nutritive values of some common Asian forages

Forage	DM (%)	ME (MJ/kg DM)	CP (%)	NDF (%)
Grasses				
Napier grass (immature)	13–15	8-9	11–15	60–70
Napier grass	18–25	7-9	10–12	65–75
Rhodes grass	20–25	7-9	8–10	65–75
Guinea grass	20–25	7-9	10–12	65–75
Para grass	25–30	7-9	10–12	65–75
Forage sorghum	13–15	7-9	11–15	65–75
Native pasture	25–35	7-9	8–10	60–70
Leucaena leaf	28–32	7-9	18–26	40–50
By-products				
Maize stover	18–25	7-8	6–8	60–70
Cassava hay	85–90	8-9	22–25	50–60
Peanut leaf and stem	26–30	9–10	18–20	50–60
Soybean leaf	26–30	10–11	20–24	50–60

Ample supplies of quality grass are the basis of profitable dairy farming (Vietnam).

Table 5.2. Nutritive value of some common Asian supplements

Feed	DM (%)	ME (MJ/kg DM)	CP (%)	NDF (%)
Energy				
Cassava chips	85–90	12–13	2–3	5–10
Maize (grain)	85–90	13–14	10–12	8–10
Rice bran (good quality)	85–90	11–12	13–14	25–35
Rice bran (poor quality)	85–90	8–10	8–10	30–50
Sweet potatoes	30–35	12–13	5–6	20–30
Cassava waste (fresh)	20–24	7–8	2–3	20–25
Wheat pollard	85–90	11–12	14–16	30–40
Brewers grain	25–30	10–11	25–30	50–60
Commercial concentrate	85–90	11–12	15–18	12–25
Protein				
Soybean meal	85–90	13–14	45–50	25–35
Coconut meal	85–90	12–13	15–25	40–50
Cottonseed meal	85–90	12–13	40–45	30–40
Palm kernel cake	85–90	11–12	14–16	60–70
Soybean curd (fresh)	13–15	13–14	20–25	25–35
Urea	100	–	280	–
Forage				
Rice straw	85–90	6–7	4–6	65–75
Urea-treated rice straw	85–90	8–9	7–8	65–75
Banana stem	5–8	9–10	3–4	60–70
Sugar cane tops	25–30	7–9	5–6	65–75

Table 5.3 provides a ready reckoner to consider when seeking feeds to balance cows' diets that may be low in dietary energy or protein.

5.3 Predicting cow performance from nutrient intakes

Cows require nutrients to survive and be productive. It is possible to describe these requirements in terms of dietary energy, protein and fibre. Through knowledge of intakes of energy and the contents of protein and fibre in the total diet, the performance of a dairy cow can be predicted.

5.3.1 Water

Lactating dairy cows in the tropics require 60–70 L of water each day for maintenance, plus an extra 4–5 L for each litre of milk produced.

Water requirements rise with air temperature. An increase of 4°C will increase water requirements by 6–7 L/day. High-yielding milking cows can drink 150–200 L water/day during the hot season. Other factors influencing water intakes include DM intake, diet

Table 5.3. Classification of supplements and basal forages according to their energy and protein contents

Energy/protein classification	Poor energy (<8 MJ/kg DM of ME)	Moderate energy (8–10 MJ/kg DM of ME)	Good energy (>10 MJ/kg DM of ME)
Poor protein (<10% CP)	Rice straw Maize stover Sugar cane tops Cassava waste	Rice bran (poor) Most grasses Sweet corn cobs Banana stem Urea-treated rice straw	Cassava chips Paddy rice Molasses Sweet potatoes Pineapple waste Maize silage
Moderate protein (10–16% CP)	–	Brown rice Well-managed grasses Soybean Immature grasses	Maize grain Sorghum grain Rice bran (good) Wheat pollard Palm kernel cake
Good protein (<16% CP)	Urea	Whole cottonseed Shrimp waste Cassava hay Most legumes Legume hays	Brewers grain Coconut meal Soybean curd Commercial concentrate Protein meals Legume leaves

Continual access to fresh drinking is essential for dairy cows (Pakistan).

Table 5.4. Energy requirements of small holder's milking cows (in MJ/day of ME) at different stages of lactation and pregnancy status

	Cow details						
Description	**Cow 1**	**Cow 2**	**Cow 3**	**Cow 4**	**Cow 5**	**Cow 6**	**Cow 7**
Live weight (kg)	550	550	550	550	500	500	500
Month of pregnancy	Empty	Empty	Empty	3rd	6th	7th	9th
Milk production (kg/day)	20	17	13	10	8	5	0
Fat test (%)	3.6	3.6	3.6	3.6	4.0	4.0	0
Protein test (%)	3.2	3.2	3.2	3.2	3.8	3.8	0
LW gain/loss (kg/day)	−0.5	0	0	0	0	+0.25	+1.0
Energy requirements (MJ/day)							
Maintenance	59	59	59	59	54	54	54
Activity	0	0	0	0	0	0	0
Pregnancy	0	0	0	0	8	10	20
Milk production	*20 × 5.1* = 102	*17 × 5.1* = 87	*13 × 5.1* = 66	*10 × 5.1* = 51	*8 × 5.5* = 44	*5 × 5.5* = 27	0
Weight gain or loss	*−0.5 × 28* = −14	*0*	*0*	*0*	*0*	*0.25 × 44* = +11	*1.0 × 55* = +55
Total energy requirements	147	146	125	110	106	102	129

composition, humidity, wind speed, water quality (sodium and sulphate levels), and the temperature and pH of the drinking water.

5.3.2 Energy

Cows need energy for maintenance, activity, pregnancy, milk production and for body condition. It is beyond the scope of this manual to describe the detailed calculations required to estimate the energy requirements for dairy cows to achieve set production targets. However, to formulate any ration for such target production levels, such calculations are necessary. For readers to follow them up, they are described in detail in my earlier book (Moran 2005).

Table 5.4 presents example of these calculations to determine energy requirements for seven different cows, with varying live weight, pregnancy status and changes in live weight, to produce target yields of milk some with different milk fat and protein concentrations. The energy requirements are expressed in MJ of ME per day.

Dairy farmers and advisers involved in formulating rations for milking cows should seek assistance to ensure the rations they plan to feed provide sufficient energy for cows to achieve production targets.

5.3.3 Protein

The amount of protein a cow needs depends on her size, growth, milk production and stage of pregnancy. However, milk production is the major influence on protein needs. Cows in early lactation require 16–18% CP in their diet and this decreases to 14–16% in mid-lactation, to 12–14% in late lactation and 10–12% during the dry period.

5.3.4 Fibre

Cows need a certain amount of fibre in their diet to ensure that the rumen functions properly and to maintain the fat test. Acceptable levels of neutral detergent fibre (NDF) in the diet are in the range of 30–35% of dry matter (DM).

5.4 The lactation cycle

Cows must calve to produce milk and the lactation cycle is the period between one calving and the next. The cycle is split into four phases: the early, mid and late lactation (each of about 120 days) and the dry period (which should last as long as 65 days). In an ideal world, cows calve every 12 months.

A number of changes occur in cows as they progress through different stages of lactation. As well as variations in milk production, there are changes in feed intake and body condition, and stage of pregnancy. Figure 5.2 presents the interrelationships between feed intake, milk yield and live weight for a Friesian cow with a 14-month inter-calving interval, hence a 360 day lactation.

Following calving, a cow may start producing 10 kg/day of milk, rising to a peak of 20 kg/day by about 7 weeks into lactation then gradually fall to 5 kg/day by the end of lactation. Although her maintenance requirements will not vary, she will need more dietary energy and protein as milk production increases then less when production declines. However, to regain body condition in late lactation, she will require additional energy.

If a cow does not conceive, she has no need for additional energy or protein. Once she becomes pregnant, she will need some extra energy and protein. However, the calf does not increase its size rapidly until the sixth month, at which time the nutrient requirement becomes significant. The calf doubles its size in the ninth month, so at that stage a considerable amount of feed is needed to sustain its growth.

Cows usually use their own body condition for about 12 weeks after calving, to provide energy in addition to that consumed. The energy released is used to produce milk, allowing them to achieve higher peak production than would be possible from

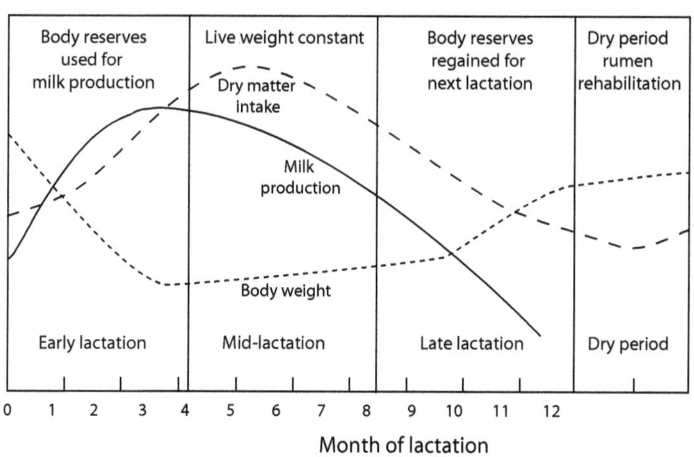

Figure 5.2. Dry matter intake, milk yield and live weight changes in a cow during her lactation cycle

their diet alone. To do this, cows must have sufficient body condition available to lose, and therefore they must have put on weight late in the previous lactation or during the dry period.

5.4.1 From calving to peak lactation

Milk yield at the peak of lactation sets up the potential milk production for the year; one extra kg per day at the peak can produce an extra 200 kg milk/cow over the entire lactation. The full lactation response to extra milk at peak yield varies greatly with feeding management during mid and late lactation. There are a number of obstacles to feeding the herd well in early lactation to maximise the peak. The foremost of these is voluntary food intake.

At calving, appetite is only about 50–70% of the maximum at peak intake. This is because during the dry period, the growing calf takes up space, reducing rumen volume and the density and size of rumen papillae is reduced. After calving, it takes time for the rumen to 'stretch' and the papillae to regrow. It is not until weeks 10–12 that appetite reaches its full potential.

5.4.2 Peak lactation to peak intake

Following peak lactation, cows' appetites gradually increase until they can consume all the nutrients required for production, provided the diet is of high quality. From Figure 5.1, cows tend to maintain weight during this stage of their lactation.

5.4.3 Mid and late lactation

Although energy required for milk production is less demanding during this period, because milk production is declining, energy is still important because of pregnancy and the need to build up body condition as an energy reserve for the next lactation. It is generally more efficient to improve the condition of the herd in late lactation rather than in the dry period.

5.4.4 Dry period

Maintaining (or increasing) body condition during the dry period is the key to ensuring cows have adequate body reserves for early lactation. If cows calve with adequate body reserves, they can cycle within 2–3 months after calving. If cows calve in poor condition, milk production suffers in early lactation because body reserves are not available to contribute energy. In fact, dietary energy can be channelled towards weight gain rather than being made available from the desired weight loss. For this reason, high feeding levels in early lactation cannot make up for poor body condition at calving.

5.5 Proportion of productive cows in the milking herd

One good measure of the performance of the milking herd is the proportion of cows actually producing milk. For herds with a 12-month calving interval, the lactation length should be 300 days (for a 65-day dry period), so the lactation length would be the calving interval minus 65 days, meaning that 82% of the cows are milking at any one time with a

Table 5.5. Percentage of days milking as influenced by calving interval and lactation length

Calving interval (days)	Dry period (days)	Lactation length (days)	% days milking
365	65	300	82
	90	275	75
	115	250	68
400	65	335	84
	100	300	75
	125	275	69
	150	250	62
450	65	385	86
	115	335	74
	150	300	67
	175	275	61
	200	250	55

100% calving rate. However, in most year-round calving systems, only a maximum of 74% of the adult cows (including the first calf heifers) are likely to be milking. The longer the dry period, the fewer the number of cows milking at any one time. The number of cows milking at any one time is influenced by several factors, the most important ones being lactation length, inter-calving interval and calving rate. Guidelines for tropical dairy farmers are:

- 74%; excellent
- 60–73%; acceptable
- 50–59%; below average
- 40–49%; not good.

One good way to demonstrate the importance of having as many of the adult cows milking as possible is to develop a table, like Table 5.5, in which the percentage of days any one cow is milking is related to the herd's inter-calving interval, the length of the dry period, hence the average lactation length. This is essentially the same as calculating the percentage of adult cows milking for a 100% calving rate.

6

Feeding management of the milking herd

This chapter summarises the key feeding practices for SHD farmers and some adverse effects of poor feeding management.

The main points in this chapter

- There are many influences on milk responses to feeding supplements.
- Cows produce less and less extra milk for each extra level of supplement fed and this is called the decreasing marginal milk response.
- Some of supplement goes immediately into milk and some goes to body fat, which produces more milk later on.
- The Asian 'rule of thumb' to feed 1 kg concentrate per 2 kg milk produced provides a safety margin to feeding management, but can increase overall feed costs.
- There are several key metabolic diseases brought about through poor feeding management, the most important being lactic acidosis or grain poisoning.
- There are many simple observations to highlight these metabolic diseases.
- Farms should not be overstocked beyond their capacity to provide sufficient home-grown quality forages.

This second nutrition chapter, on feeding management, discusses milk responses to supplements, problems with feeding unbalanced diets and optimum stocking capacity on small holder farms. A four-page advisory leaflet, originally developed for Indonesian SHD farmers, on guidelines for feeding and herd management is presented in Appendix 4.

Because tropical forages can only support low levels of milk production (only 6–8 kg/cow/day), they must be supplemented with high energy concentrates to achieve target milk yields. If dietary protein is in short supply, high protein supplements should also be fed.

Supplements are fed to improve or maintain milk production, cow condition or reduce intakes of basal forages when in short supply. Basal forages are the major forages fed by SHD farmers, while supplements include all the additional feeds offered to

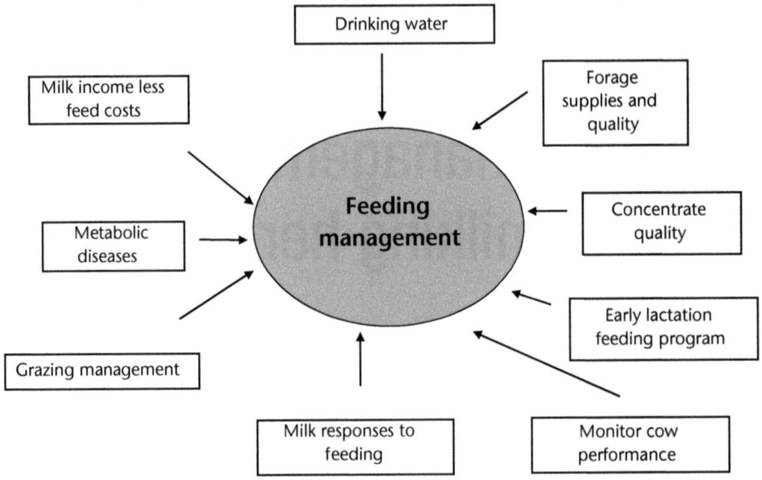

Figure 6.1. The key factors to consider when planning the milking herd's feeding management

improve cow performance. Where milking cows can graze for much of the day, the basal forage would be pasture.

Most SHD feeding systems are then based on hand feeding a single or combination of forages together with supplements. Farmers usually feed a set level of forage and increasing quantities of supplements depending on target milk yields.

The impact of supplementary feeding is difficult to assess because the results may not appear immediately as milk. In the short term, the response to a particular supplement may be small. But if that supplement is used to 'save' on other feeds – for example, until the basal forages become more readily available or until a lower-priced supplement (such as a crop residue) can be used – then it may be important and ultimately profitable.

The factors affecting responses to supplementary feeding are numerous (see Figure 6.2) and their interactions are complex.

6.1 How cows respond to supplements

6.1.1 Deceasing marginal responses

As the intake of energy increases, the amount of extra milk produced from each extra unit of energy decreases. In other words, the marginal, or additional, milk response decreases as the level of supplement intake increases.

The major reason for this decreasing marginal milk response is that, with successive increments of feed energy, the cow increasingly partitions nutrients from milk production towards body tissue deposition as milk production approaches the cow's genetic potential. In addition, the stage of lactation has an influence on how much of the supplement's nutrients 'go into the bucket' and how much 'go on the back'. Modern day milking cows in early lactation tend to lose weight to divert additional nutrients towards milk, while those in late lactation tend to repartition nutrients to replace previously lost body reserves.

This is how much fresh grass a milking cow can eat in one day.

Gliricidea is a high-quality tree legume suitable for milking cows (Sri Lanka).

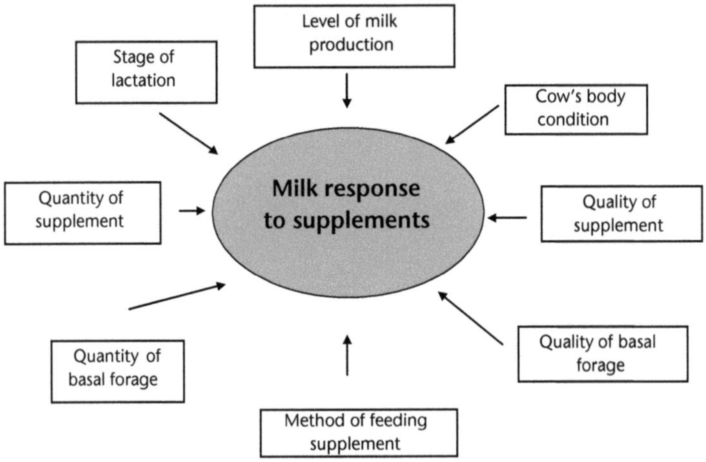

Figure 6.2. The key factors influencing milk responses to supplements

A second reason for declining marginal responses is that utilisation of one feed type can change with increasing intake of a second feed type, which is known as an associative effect. Efficient digestion of forages, particularly low-quality forages, requires an adequate population of fibre-digesting microbes in the rumen. By feeding increasing amounts of high-starch concentrates, the proportion of these microbes will decrease as more starch digesting microbes propagate as a result of the higher starch intake. Consequently, the digestion of the forage can decrease with increasing intakes of such concentrates. Additional starch excretion may also occur, further reducing feed utilisation. This can be particularly important when feeding high levels of supplements rich in fermentable carbohydrates, because rumen pH can decrease, dramatically reducing fibre digestion.

Supplementary feeding usually results in higher total feed intakes. Increasing intakes are the result of decreased times that consumed feed spends in the rumen where it is exposed to microbial breakdown. If less of that feed is digested and the nutrients are absorbed into the blood stream or pass down the digestive tract, less dietary energy becomes available for use by the animal. The cow partly compensates for this through decreased losses of energy via methane and urine with increasing feed intake. Although this may not be important unless total feed intakes dramatically increase through supplementation, it can contribute to declining marginal milk responses to supplements.

Another factor decreasing milk responses is the often incorrect assumption that all of the supplement is actually consumed. Rarely is there nil wastage, particularly if the supplement is roughage. Fortunately, stall feeding minimises such wastage, compared with feeding cows while outdoors.

The major difficulty when predicting milk responses to supplementation, even if substitution rates are known, is the lack of information on the relative importance of the above factors. Without such knowledge, dairy advisers can only, and probably incorrectly, assume additive effects when feeding a mixture of various feed types, which would tend to overestimate such milk responses particularly when there are marked differences between basal roughages and supplement type or large amounts (say 5 kg DM/cow/day or more) of supplement are fed.

6.1.2 Immediate and delayed milk responses

Responses to supplementary feeding have both immediate and delayed components. Some of the supplement goes immediately to milk production and some goes to body fat, which contributes to milk production at a later stage when this condition is mobilised.

To manage the feeding of supplements effectively, it is important to know how cows respond to them. The response is variable and depends on the circumstances in which the supplement is fed.

The response in milk yield is generally due to the extra energy in the supplement. Unless the supplement improves the use of nutrients already in the diet or stimulates intake of the basal forage, farmers will not get any more milk than that produced from the energy the supplement contains.

In practice, forage substitution almost always occurs, resulting in the response being less than that predicted from the amount of energy in the supplement. The response will reduce at least by the equivalent of the energy in the forage no longer eaten. Also, some of the energy in the supplement goes to condition score rather than directly into milk, so the immediate milk response will be even smaller. Most experiments have only measured the immediate response to supplements. Because they are short term (usually only several weeks), they cannot measure the delayed milk response from body condition, hence the total milk response.

We know most about the immediate response to supplements from studies in temperate countries. Whether these will be similar to responses in tropical countries requires further research. The major differences between temperate and tropical climate zones is the poorer quality of tropical forages and the fact that many supplements are based on by-products, which vary greatly in nutritive value in tropical countries. Another difference may be the poorer quality control in feed mills, hence the greater variation in energy and protein contents of formulated concentrates in tropical countries. Therefore, it is highly likely that milk responses in Asia will be lower than those in temperate countries.

6.1.3 Guidelines for temperate grazing dairy systems

In early lactation, the average immediate response to feeding concentrates containing 12 MJ/kg DM of ME is 0.6 kg of milk per kg of supplement DM, ranging from 0.2 to 1.0 kg.

In mid-lactation to late lactation, the average immediate response is 0.5 kg of milk per kg of supplement DM, ranging from 0.3 to 0.8 kg.

One generalisation sometimes made is that 'you get half the response now and the other half later, when the condition score energy is converted back to milk'.

6.2 Milk:concentrate ratios in production rations

Many Asian dairy advisers use a general 'rule of thumb' that for every 2 kg of milk produced above that supplied from forages, farmers should feed 1 kg concentrate. This is a safety measure because of lack of knowledge on the nutritive value of the feeds, particularly the forages. It also provides supplemental energy to cows when fed only limited amounts of forage. In any dairy system, whether in temperate grazing systems or Asian SHD systems, the principles for feeding milking cows should be:

1. feed sufficient quality forages first, then
2. supplement with concentrates, which are
3. formulated to overcoming specific nutrient deficiencies
4. to achieve target milk yields.

With knowledge of the feeding value of the forages and concentrates, and their costs, more objective, hence better, decisions can be made on how much concentrates should be fed to achieve target milk yields. Granted this requires more knowledge and greater effort than following the 'feed 1 kg concentrate per 2 kg milk' rule, but such decisions can greatly reduce feed costs, hence improve profitability, when expressed as milk income less feed costs.

Table 6.1 presents a series of milk:concentrate ratios to achieve target milk yields. When cows are fed better quality forages, more milk is produced per kg concentrate fed. The 2:1 (1 kg concentrate/2 kg milk) rule is only applicable with very low-quality forages, namely those with ME contents of 7–8 MJ/kg DM.

Milk production is very responsive to nutrient intake. Among livestock producers, dairy farmers are very fortunate in that their cows' milk yield today is directly affected by their feeding management yesterday. No other type of livestock provides such a rapid feedback to herd management. Once farmers set their target milk yields, as long as they are realistic about their farming system, they can monitor their success or failure in achieving these by gradually changing one of the feeds in the cows' ration. If the additional milk produced, as feeding levels are improved, returns more than the additional feed inputs, then that was a profitable management decision.

Farmers should change their feeding program, but only one feed at a time, say once each week, then note the milk response. They should also note changes in other feed inputs. For example, if they increase concentrates and find cows eat less forage, and know the cost of energy or protein in the various feeds, they can then decide on the most profitable combination of these feeds.

Table 6.1. Milk:concentrate ratios (kg milk produced/kg concentrate fed) to achieve target milk yield in cows fed forages of varying quality

Milk yield (kg/day)	Forage quality (MJ of ME/kg DM)			
	7.3	8.2	9.0	9.9
6	1.8	8.6	–	–
10	2.0	4.0	12.5	–
14	2.1	2.9	12.7	46.7
18	2.2	3.0	6.0	25.7
22	2.2	2.9	4.1	12.9

6.3 Problems with unbalanced diets

Diets should be properly balanced for energy, protein, fibre and certain minerals to ensure optimum cow performance. This section discusses some of the indicators of unbalanced diets and also the major metabolic disorders that can be traced back to nutrient deficiencies.

6.3.1 Some indicators of unbalanced diets

The most important indicators of unbalanced diets can be used to identify dietary problems, these being:

- **Lack of rumination.** After an initial period of eating forages, animals normally start to ruminate or chew their cud. If this is not occurring in much of the herd (say 50%), then there may be a lack of fibre in the diet. This may be confirmed by looking for changes in milk composition as described below.
- **Loose manure.** If faecal material is very loose and watery, it may indicate a lack of fibre in the diet. Assessing any changes in milk composition can also check this.
- **Low milk fat test.** A drop in the milk fat test tends to occur when the herd is placed on a low-fibre diet (such as a diet high in cereal grain and very immature forage). The easiest way to increase the fibre content of the diet is to feed hay or straw. Care needs to be taken though when feeding out poor-quality forages. A drop in dietary energy intake could cause milk and protein yield to fall.
- **Low milk protein (or solids-not-fat) test.** Low milk protein or solids-not-fat (SNF) content is common in early lactation when cows are in negative energy balance. In other words, their energy needs are greater than their intakes, causing them to lose body condition. Shortages of energy reduce protein utilisation by rumen microbes. As a result, the supply of microbial protein – the cows' major protein source – is reduced. Under most circumstances, providing a higher energy diet will lift protein or SNF. Cows will only respond to protein supplementation with a lift in protein or SNF if they are truly deficient in dietary protein. This is because they are unable to use energy properly when there is a protein shortfall.
- **Reduced feed intake.** Many of the causes are discussed in the following section.

6.3.2 Metabolic disorders and unbalanced diets

Metabolic disorders can be clinical, when there are obvious symptoms, or sub-clinical, when there are not. Even at the sub-clinical level, they can depress feed intake and cause production losses.

Metabolic disorders such as ketosis and acidosis are usually linked to low intakes around calving or abrupt changes in diet.

Managing nutrition well during the dry period and in early lactation is the key to preventing or minimising the occurrence of metabolic disorders. The aim is to:

- maximise nutrient intake around calving and in early lactation by providing enough high-quality feed
- avoid decreases in intake caused by sudden changes in diet when cows calve and join the milking herd.

Table 6.2. Metabolic disorders arising from unbalanced diets

	Cause	Symptoms	Treatment
Milk fever	Sudden decrease in blood calcium levels	Decreased intake and milk yield. Calving paralyses, death.	Feeding management prior to calving to stimulate cow's ability to mobilise body calcium
Grass tetany	Low blood magnesium levels	Decreased intake and milk yield. Muscular staggers, death.	Feed magnesium supplements
Ketosis or acetonaemia	Cows rely on fat reserves for energy during early lactation	Decreased intake and milk yield, characteristic smell of breath	Feed well-balanced diet during early lactation
Lactic acidosis (grain poisoning) and laminitis	Rumen pH becomes very low due to high starch intake	Decreased intake and milk yield. See below.	Include rumen buffers in diet and sufficient roughage
Bloat	Build up of foam in rumen, which stops gas from escaping	Left side of cow is swollen. Cow stands up and lies down frequently.	Put hose down oesophagus, administer oil, stab left flank to release gas
Urea toxicity	Ammonia poisoning	Rumen stops moving, death	
Feed toxicities	Anti-nutritional factors in diet. See below.	Sickness and death	Identify cause and remove from diet

The most important metabolic disorders are summarised in Table 6.2, while one of the most prevalent disorder, lactic acidosis, is discussed in more detail below.

6.3.3 Lactic acidosis

Acidosis can be clinical (with cows obviously sick) when the rumen pH falls below 5.0 or sub-clinical when the rumen pH falls below 5.5. Symptoms of sub-clinical acidosis include:

- low milk fat, below 3.0–3.3%
- low milk protein
- reduced milk yield
- reduced feed efficiency
- sore feet due to laminitis or overgrown claws
- manure in cows on same diet varying from firm to very liquid
- manure foamy, containing gas bubbles
- manure containing larger than normal lengths of undigested fibre, more than 1.2 cm long
- manure containing undigested yet ground grain, less than 3.5 cm in size
- limited rumination, less than 50% of the cows cud chewing while resting
- cyclical feed intake.

To avoid acidosis, high-starch feeds should be introduced gradually (i.e. 0.5 kg grain/cow/day) so that the population of rumen microbes can adjust according to the type of fermentation that is required (more starch-fermenting microbes may be needed). Different cows respond differently to grain feeding. Some cows can handle 6 kg grain/day, while others will get sick on 3 kg/day; there is always a cow that will eat more than her fair share. The key to success is to make it a gradual daily increase and to watch the cows and check for symptoms of acidosis or grain poisoning.

Acidosis can be overcome by feeding more fibrous roughages, but that can lead to reduced feed intake, hence milk yields. Buffers can be included in the diet to stabilise rumen pH so that the rumen environment allows a healthy population of rumen microbes.

Feeding management can also influence the incidence of sub-clinical acidosis in that when cows cannot eat when they are hungry, they overeat (having a larger than normal feed) when they eventually get access to the feed trough. In this case, the acidosis is not caused by lactic acid, but by excess production of the volatile fatty acids from rumen digestion. It is important therefore that all cows should be able to eat when they want to.

Farms where cows are less able to lie down, hence they spend too long standing, particularly on hard surfaces, can have greater problems with sore feet due to both trauma and acidosis. Cows should be able to lie down for at least 8 hr each day. Other factors that can increase problems with sore feet include:

- heat stress (when some cows prefer to stand)
- cows spending too long waiting to be machine milked
- cows with 'perching' behaviour: namely standing with their front feet in the feed trough and their back feet on the floor.

6.3.4 Feed toxicities

Not all animal health problems arise from unbalanced diets. Some feeds contain anti-nutritional factors such as:

- prussic acid (cyanide) in some varieties of forage sorghums, when harvested or grazed as immature crops, or when drought stressed
- hydrocyanic acid in fresh cassava forage
- mimosine in some varieties of *Leucaena*
- gossypol in whole cottonseed and cottonseed meal
- tannins in banana stems and leaves, mango seed kernel or sal seed meal
- trypsin inhibitor in soybean products.

Mouldy silage can contain bacteria such as *Listeria*, which can cause abortions in dairy cows and can also pass through the milk and infect humans. Noxious fungi in silage can lead to pneumonia and abortions in cows. Many different coloured moulds can be found in poorly preserved silages, but unfortunately their colour is not a good guide to the type of mould or its toxicity.

The high humidity and temperature of tropical environments encourages the growth of many contaminating microbes. For example, mycotoxins and aflatoxins such as *Fusarium* and *Aspergillus* can grow on moist cereal grains and by-products. Feed ingredients, of both plant and animal origin, are frequently contaminated with *Salmonella*, which can cause disease and death, particularly in young calves.

In humid tropical countries, veterinary drugs may be administered in animal feeds, some of which have been banned in many Western countries. There can also be potential hazards, to both animal and human health, from excessive levels of herbicides, pesticides and fungicides, or other industrial/environmental contaminants, such as heavy metals. Many of these can accumulate in animal tissues and be excreted in milk.

The recent human health issues arising from 'mad cow disease' (bovine spongiform encephalopathy) have led to bans in most countries in feeding products derived from ruminant animals back to ruminants.

6.3.5 Other feed additives

A feed additive can be described as a feed ingredient that produces a desirable animal response.

Feed additives have gained attention and use in recent years. Expected responses from feed additives include higher milk yields, increases in milk fat and protein contents, improved DM intake, a more stable rumen pH and improved fibre digestion. It must be stressed that these additives may be more suited to intensive production rather than small holder dairy systems.

The primary feed additives currently being used are ionophores and antibiotics. Monensin (sold as Rumensin®) and lasolocid (sold as Eskalin®) are two commonly used additives that produce their effects by modifying the rumen environment. They alter the microbial population of the rumen, which in turn changes the mix of end products from microbial fermentation.

Rumensin® reduces the population of microbes that produce methane gas (which cannot be used by the cow as an energy source). The proportion of microbes that ferment feed to other more useful sources of energy is increased, resulting in improved milk yields. Responses to this additive depend on the diet and the stage of lactation.

Eskalin® inhibits the microbes that produce lactic acid and can therefore play a role in preventing lactic acidosis.

6.3.6 Troubleshooting feeding problems

There are many simple observations farmers can use to highlight problems with feeding management. Such quick checks include:

- manure consistency, colour and content (see Table 6.3 below)
- cows are actually eating all the concentrates on offer
- rumination; ideally 50% of the herd should be ruminating when resting
- hair coat; appearance and cleanliness
- cow visual appearance; the diet should be reviewed if the cows are looking poor with dull sunken eyes, scruffy coat and hunched backs
- respiration rate, coughing and nasal discharges
- mobility of legs and feet
- body condition at different stages of lactation
- physical appearance and smell of forages
- physical appearance and smell of concentrates
- sudden changes in milk yield
- sudden changes in milk composition, namely fat and protein (or solids-not-fat) contents

- metabolic problems, as discussed above
- physical conditions in shed such as cleanliness and ventilation.

Keep in mind that a sudden change in one of these quick checks may be due to a temporary fluctuation in nutrition. Provided that the check quickly returns to normal, cow performance may not be adversely affected. It is important to take immediate action when a quick check remains abnormal for several consecutive days and/or several quick checks become abnormal at the same time.

6.3.7 Monitoring manure consistency

Manure that is excessively loose or dry and firm for the diet fed may indicate a dietary imbalance that requires action. Manure pats can be easily evaluated using a 1 to 5 scoring system as described in Table 6.3.

When assessing the herd's average manure score, it is important to assess the range in consistency of faecal pats. If more than 20% are one score or greater (or less) than the average, this may indicate a nutritional imbalance or management problem. With cows living in sheds, the consistency of manure should only be made on faecal pats on clean floors, that is without any urine contamination that will change the consistency over time.

6.4 Optimum stocking capacity

Forages always provide a cheaper source of the key feed nutrients (energy and protein) than do concentrates. It is usually cheaper to grow these forages on the farm rather than purchase them. It is easier to control forage quality on farm, through fertiliser and harvest interval, than with purchased forages. When relying on off-farm forage supplies,

Table 6.3. Evaluating manure consistency

Score	Manure description	Action required
1	Very liquid manure with consistency of pea soup. May leave cow's rectum in a steady flow. Includes cows with diarrhoea.	Increase effective fibre intake and seek nutritional advice
2	Runny manure which does not form a distinct plie. Manure will splatter on impact and form lose piles less than 25 mm high.	Consider increasing effective fibre intake
3	Manure has porridge-like consistency. Forms a soft pile 40–50 mm high, which may have several concentric rings and a small depression in the middle. Make a plopping sound when it hits concrete floors and will stick to the toes of your shoes.	This is the desired consistency
4	Thick manure, sticks to shoes and readily forms piles more than 50 mm high	Consider reducing effective fibre intake and increasing concentrate intakes. Seek nutritional advice.
5	Manure appears as firm faecal balls	Consider reducing effective fibre intake and ensure adequate drinking water is available. Seek nutritional advice.

Table 6.4. Optimum stocking capacities for small holder dairy farms with different levels of forage management (Moran 2005)

Quality of forage management	Poor	Average	Good
Forage yield:			
t DM/ha/yr	10	20	30
t fresh/ha/yr	67	130	200
Milking units/ha forage	3.4	6.9	10.3
Adult cows/ha forage	4.0	8.1	12.1

One milking unit is one adult cow plus 20% of a replacement heifer.
Assumed forage intakes:
7.5 kg DM/day for 275 day/yr for milking cows
4.5 kg DM/day for 90 day/yr for dry cows
3.0 kg DM/day for 365 day/yr for 20% of a replacement heifer.

farmers depend on what is available, either from traders who harvest the roadsides, paddy fields, tree plantations or forests or from other farmers who sell their excess supplies, either as crop by-products (such as rice straw or corn stover) or forage crops specifically grown for sale.

In my earlier book (Moran 2005), I listed a series of assumptions and calculations of optimum stocking capacities for SHD farmers with different level of forage management. Calculations were made for farmers who run replacement heifers on the same farm as their milking herd and for farmers that have them reared off farm. The calculations also included three levels of forage management – namely poor, average and good – to produce 10, 20 and 30 T forage DM/ha/yr respectively. Table 6.4 presents the range of optimum stocking capacities.

For a farmer growing the maximum quantities of quality forages, to feed his milking cows well, he should have no more than 8–10 milking cows per ha of forage grown on his farm. However, most dairy small holders do not manage their forages well enough to produce the highest yields of forage. Therefore a more realistic recommendation would be 6–8 milking cows (plus the replacement heifers) per ha of forage grown on farm.

Unfortunately, most SHD farmers like to keep more cows than this recommendation, meaning they must either have to purchase forages off farm, underfeed their milking cows (and heifers) with less forage, or if they aim to produce high yields of milk (say more than 12–14 kg/day), feed excessive levels of concentrates to each milking cow. This is a more expensive way to produce milk, and frequently leads to digestive problems, such as sub-clinical acidosis. Not only will this reduce feed efficiency, it will increase the cost of production and reduce farm returns. Therefore, as all good business owners aim to, farmers should produce at the optimum level to maximise efficiencies and profits. Farmer shouldn't put too many cows onto their farms if they cannot feed and manage them properly.

7

Disease prevention and control

This chapter discusses the most important diseases encountered by SHD farmers in the tropics, their treatment and some of the prevention measures.

The main points in this chapter

- Farmers can develop observation skills to assess the general health status of their herd and individual cows.
- There is a wide diversity of disease agents, ranging from parasites to microbial agents to unbalanced nutrition.
- Every farmer should develop a herd health plan that includes a good recording system and plans for introducing new stock onto the farm.
- Poor animal health can have adverse effects on reproductive performance.
- Cow lameness is a common problem on small holder farms where cows are continually tethered.
- Mastitis has many hidden effects on cow performance and farm profits as sub-clinical infections cannot be readily identified and treated.

It is not easy to say what 'disease' means. In a general sense it means any aspect of a cow's health that is not 'normal'. Disease is then a condition that is detrimental to the health and wellbeing of that animal. It can include injuries, infections by micro-organisms, infestations of parasites, nutritional deficiencies, poisoning and hereditary abnormalities. The presence of disease, whether acute or chronic, reduces cow performance, which invariably reduces farm production and profitability.

7.1 Physical attributes of healthy and sick cows

The physical state of a cow is a good guide to its health status. Healthy animals are alert and active, and have bright eyes, with no discharge, smooth and shiny skin, they breathe and urinate regularly and their tail moves to drive flies away. Signs of stress include loss of appetite, reduced daily milk yield, increased temperature, high respiratory rate,

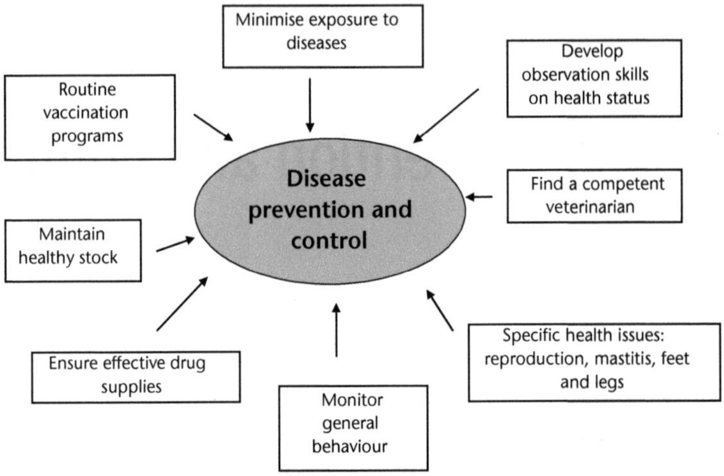

Figure 7.1. The basic elements of the milking herd's disease prevention and control program

protruding tongue, breathing with an open mouth, inability to lie down and a preference for remaining in cool waters, such as swamps.

Symptoms of health problems range from one almost unnoticeable changes in behaviour to sudden death. Some of the signs to look for include:

1. **Nutritional status.** Cows can be fat, normal or thin. Thin cows are not necessarily sick, because they could have recently calved or simply be high-producing animals. However, sick cows tend to lose weight due to depressed appetite, poor feed digestion or loss of body reserves. Cow condition should then be judged in relation to all circumstances.
2. **Walking and standing.** The way an animal moves can indicate pain in the body, the result of a traumatic injury or an infected hoof. A method of scoring the cow's ability to walk (the locomotion score), which provides a good guide to lameness, is described later in this chapter.
3. **Eyes and ears.** Eyes should have a bright and lively expression with no discharge. Sunken eyes indicate dehydration. Ears should be able to freely move around.
4. **Skin, coat or mucous membranes.** The skin of healthy cows is flexible and, when pinched, should quickly return to normal; a lengthy delay indicates dehydration, as does a dry nose. The coat should be smooth and shiny. The mucous membranes around the eye, nose and vagina should be pink to reddish in colour and be moist. In sick cows, these membranes can become either too red or too pale, the later indicating anaemia.
5. **Digestion.** Healthy cows have a good appetite and eat with eagerness. Faeces and urine are discharged regularly, with the faeces having a normal consistency. When digestion is disturbed, the cow's appetite decreases and the faeces is discharged too fast (scours) or too slow (constipation). Cows ruminate frequently when healthy (at least 6–8 hr each day), and if she does not ruminate when resting, her digestion is

disturbed. Troubleshooting problems with feeding management are discussed in Chapter 6 of this manual.

6. **Urine.** It should be thin, yellow and clear; thick, mucous or red urine is an indication of ill health.
7. **Vagina.** It should be closed, with no swelling, no discharge and be slightly whitish-red in colour; a swollen vagina with whitish discharge or decomposing membranes is indicative of reproductive problems.
8. **Respiration.** In healthy cows, respiration is quiet and regular, whereas in cases of unrest, fever, fatigue or heat stress, respiration rates increase. See Chapter 12 for a guide to heat stress based on respiration rates. Coughing, nasal discharge, and rapid or slow breathing can all be symptoms of ill health.
9. **Blood circulation.** Pulse rate is a good guide to cow health, with 60 to 70 beats/min being normal for dairy cows.
10. **Body temperature.** Normal body temperature is 38.5–39.5°C. Higher temperatures are recorded in sick animals (with a fever) or in heat-stressed animals.
11. **Milk production.** When a cow is sick, milk production drops, due primarily to decreased appetite.
12. **Specific signs to look for in calves.** Health problems are likely in calves with droopy ears, head down, not drinking, lying in a corner, dribbling, limping, swollen joints, swollen navel, scour or blood in their faeces. The nose and eyes should be clear and damp with no discharge. Calves should stretch when they stand up following a rest period. An odour of ammonia can indicate poor ventilation and potential pneumonia problems.

7.1.1 Other tools to diagnose diseases

There is much that farmers can do to assist veterinarians or other animal health professionals in diagnosing and treating diseases. These include:

- Examine the history. For how long has the animal been sick, how rapidly did the sickness develop, what are the breeding and calving records of the animal, have new stock recently been introduced to the farm, or have there been any recent environmental changes (such as hot weather)?
- Examine the animal. Use the checklist above.
- Examine the environment. Are other animals sick, could there be poisons or mineral deficiencies involved, what is the water quantity and quality, what are the physical conditions, such as lanes or yards, that cause traumatic problems?
- Prepare for the veterinary visit. Record the important major symptoms using the above checklists, have the sick animal easily accessible with suitable restraining equipment available, listen carefully and take notes for follow up treatment.
- Once the disease has been diagnosed and its cause, symptoms and treatment identified, it is important to develop a control program to reduce its incidence in the future. Veterinarians can assist with such a control program. For the farmer, prevention is just as, and often more important than, cure.

7.2 Disease agents

This section comprises detailed lists of the many agents that cause disease in tropical dairy cattle. They are separated into two main categories: parasites and microbial agents, followed by nutritional (and metabolic) diseases, then miscellaneous agents.

7.2.1 Parasites

Parasites that infect cows include:

- **ectoparasites** (live outside body): ticks, flies, lice, mange, fungus (e.g. ringworm), bacteria (pink eye), viruses (warts)
- **endoparasites** (live inside body): intestinal worms, lung worm, liver fluke
- **tick and insect-borne diseases:** theileriosis (East coast fever), babesiosis, anaplasmosis, bovine ephemeral fever (three day sickness), trypanosomiasis.

7.2.2 Microbial agents

Microbial agents that infect cows include:

- **bacteria:** brucellosis, vibriosis, tuberculosis, paratuberculosis (Johne's disease), pneumonia (also virus agents), trichomoniasis, campylobacteriosis (or vibriosis), leptospirosis, anthrax, and tetanus, black leg and other clostridial diseases, haemorrhagic septicaemia (HS), mastitis, infectious foot rot, joint ill
- **viruses:** rinderpest (cattle plague), foot and mouth disease, enzootic bovine leucosis, bovine viral diarrhoea, bluetongue, infectious bovine rhinotracheitis, pneumonia
- **others:** bovine spongiform encephalopathy (or mad cow disease), facial eczema (fungal), ergot poisoning (fungal), blue green algae poisoning, eye cancer
- *E. coli*, cryptosporidia, rotavirus, coronavirus, *Salmonella*, coccidia (causing **calf scours**).

7.2.3 Nutritional and metabolic diseases

Nutritional and metabolic diseases that affect cows (see also Chapter 6) include: acidosis, milk fever (low blood calcium), grass tetany (low blood magnesium), acetonaemia (or ketosis), bloat, displaced abomasum, photosensitisation, urea toxicity, nitrite toxicity, lead poisoning, foreign bodies (hardware disease) and miscellaneous plant poisonings (e.g. bracken fern, lantana, oleander).

7.3 Managing dairy herd health

The operation of a dairy farm for maximum profit includes good feeding, breeding, care and management, as well as good record keeping and a dairy health program. The objectives of a health program are to prevent the introduction or occurrence of disease and control the spread of infectious diseases and parasites.

The major role of government is to eradicate epidemic diseases and to minimise the transmission of diseases from dairy cattle and dairy products to humans. The role of dairy cooperatives is to encourage health care and eliminate diseases transmitted from cattle to humans and those that adversely influence milk spoilage.

The roles of veterinarians and dairy extension staff are to emphasise continuing education for farmers in areas of farm management and disease control, as well as providing effective services. In the final analysis, it is the small holder who must understand the reasons for good management practices and the links between nutrition, housing and health.

A written herd health plan should be developed that contains management actions for herd health to prevent and control diseases and disorders that are specific for the region. The plan should contain actions for foot care, mastitis, parasite control, infectious diseases and calf care. As part of this herd health plan, records should be kept of health and other disorders in order to identify changes and assist in management. For example, reproductive failure, udder disorders, low production, calving associated disorders, lameness and mastitis. This plan should also include biosecurity measures for buying new stock and for visitors.

7.3.1 The importance of good record keeping

An important aid for farm management is the keeping of records of all animals and events relating to animals throughout their lives. In some countries, record keeping is provided, supported and designed by government or dairy cooperatives. Records can either be electronic (using computer technology) or be kept as individual cow cards at farms or at the dairy health service or artificial insemination service centre. Records of insemination, birth date, sire, dam, calving date, vaccination date, health problems, treatment, milk yield and feeding can help farmers to predict future or preventative needs for health care.

These records also provide beneficial and relevant information for veterinarians to make correct diagnoses. Therefore, it is best to have well-organised records kept for each animal, with the forms designed to allow for easy interpretation. Most small holders do not seem to pay sufficient attention to keeping good records.

Large-scale dairy herd improvement programs are beneficial tools for farm management but the procedures tend to be costly, and need a lot of technicians to access small holders in the tropics together with considerable education for farmers to use and keep using the data. It is better to start with simple systems in milk collection centres rather than work towards a national database, such as is found in most Western dairy industries.

7.3.2 Nursing and support

Once the veterinarian has provided treatment, it is important to continue to nurse sick stock. Supportive care that can be provided includes:

- providing good nutrition to treated stock, with freely available water and quality palatable forages and concentrates
- ensuring access to water, shade and clean bedding

- maintaining a low-stress environment, with adequate housing, removing competition from other stock, removing parasites, providing pain relief, and cleaning and dressing any wounds
- taking special care of stock unable to stand, providing support such as small hay bales to stop them rolling onto their side, moving them to dry, warm shelter with good footing, in case they try to stand.

7.3.3 Scours

The state of manure is a good indicator of cow diet and general health. Scours (or diarrhoea) in adult cows can have many causes, with dietary scours the most common. Other causes of scours are acidosis (or grain poisoning), internal parasites, bacterial infections (such as *Salmonella* and Johne's disease) and viral infections (such as bovine viral diarrhoea and malignant catarrhal fever). The manure may be watery and range from green to light yellow and may or may not contain blood or pieces of bowel, or whole or poorly digested grain. Chapter 4 provides details about scours in milk-fed calves.

For acute scours, antibiotics may be needed in the case of bacterial scours, or rumen buffers (such as sodium bicarbonate) for acidosis. Providing a palatable effective fibre source can also aid in cases of digestive disturbances. For chronic scours, a worm drench may suffice. A scoring system for manure consistency is described in Chapter 6.

7.3.4 Responsible drug handling

Responsible drug handling and administration are the key to a successful animal health program. Drugs should be sourced from veterinarians or registered agricultural merchants because medicines obtained from other sources may not be safe or effective. They should be stored correctly in accordance with the instructions on the label. Storage temperature is critical for some medicines, especially vaccines. Light can damage others. Drugs should be stored securely and locked where practicable and kept out of reach of children, animals and anybody not supposed to handle them.

Keep records of:

- how much was purchased and when
- the batch number and expiry date
- when it was used and on what type of stock
- the withholding period (start and end dates) for sale of milk or for slaughter.

Use drugs only on animals recommended on the label. Dispose of unused medicines safely when treatment is finished. If using disposable needles and syringes, dispose of them after use in a safe container. For other reusable equipment, clean and sterilise before and after use. If using a syringe that requires filling from a bottle between doses, use one sterile needle left in the bottle during use to fill the syringe and a separate needle to inject the animal. Use a separate needle, or at least sterilise it, between animals. Make sure the injection site is through an area of clean and dry skin.

High standards of sanitation are required at all times to prevent rapid spread of infectious diseases in both young stock and milking cows. The most effective way of

Veterinary drugs should be stored in a cupboard away from the sun (Sri Lanka).

destroying disease-carrying micro-organisms is cleaning and disinfecting (sterilising or sanitising). However, the latter has little effect unless the surface is first cleaned. The best cleaner and disinfectant depends on the type of surface.

7.3.5 Biosecurity when purchasing new stock

Most dairy farmers purchase new stock, so it is important to plan their introduction to minimise the risk that they will introduce infectious diseases. Three factors are important in reducing this risk:

1. protecting the herd with proper vaccination
2. the source of purchased stock, including how they are transported to the farm
3. the method used to introduce the new stock to the rest of the herd.

When purchasing new stock:

- ensure their health status is known
- where possible, ensure details of their vaccination program are known
- avoid purchasing stock from unknown sources or stock that have mixed with other cattle before sale
- purchase heifers because they can be more easily quarantined and are less likely to have mastitis

- calves purchased with no signs of disease should be kept separate for at least a week
- transport purchased cattle preferably in the farmer's vehicle, in a clean truck or trailer.

Other steps that can be taken with newly purchased stock include quarantining them in an area separate from other cattle on the farm, using a medicated foot bath before allowing them to enter the herd, and vaccinating them during the quarantine period to make sure they are integrated into the farm vaccination program.

Other biosecurity measures that should be considered include controlling the movement of people, animals and equipment onto the farm. Some diseases are spread on clothing and boots. If equipment is borrowed from other farms, it should be cleaned prior to use. Keeping transport or service personnel away from the main herd area, especially the calf shed, or at least providing them with footwear, will also reduce the likelihood of introduced diseases.

7.3.6 Animal and human health

Government veterinary services must maintain surveillance of infectious or notifiable diseases, such as rinderpest, foot and mouth disease and contagious pneumonia, through vaccination and quarantine measures. However, as farms become more intensively managed, non-infectious diseases play a more important role in limiting cow performance. These could be called disease-causing risk factors such as undernutrition, poor hygiene and other management factors affecting herd productivity.

Farmers are only interested in herd health programs when the link with production is clear, such as declining milk yields, increasing mortality and poor reproduction. Mastitis, for example, is hard to manage because the sub-clinical form is very prevalent, difficult to detect and causes higher milk losses per affected cow. It is often difficult to incorporate economic parameters in such programs because of the small number of stock, and hence the influential impact of the performance of each animal.

There are many diseases that can be transferred from intensively managed livestock to humans. These include *Salmonella* (and other calf scour-causing micro-organisms), ringworm, mange (and other skin diseases) and *Leptospirosis*. Children are particularly susceptible because of their affinity to young calves, and their poor understanding of human hygiene. Other potential hazard for young children are veterinary drugs and chemicals used for cleaning or sanitation. These should be stored in a secure place.

7.4 Health and reproductive performance

There are many health management problems that can adversely affect fertility, such as:

- management of twin calves
- assisted calving
- retained foetal membranes (RFMs)
- uterine infections and vaginal discharges
- lameness
- ketosis
- displaced abomasum

- cystic ovaries
- abortions.

Some health problems affect the reproductive tract directly (such as RFMs and vaginal discharges) while others reduce feed intake leading to rapid body condition loss and anoestrus (such as lameness and ketosis). It is important to keep good records and have a planned approach for treatment and prevention.

Cows with problems at calving have an increased risk of infection of the reproductive tract. Such infections can last for weeks (even months) after calving and cows can even show normal heats with no abnormal discharge, yet still have reduced fertility. They may cure themselves over time but are more likely to show repeat heat cycles. As well as immediate treatment, follow up treatments are available such as prostaglandins or antibiotics.

To improve cow health, it is important to keep accurate records and seek veterinary advice if the percentage of naturally calving cows (excluding cows induced to calve early) with this problem exceed the number shown in brackets below:

- any cow having twins (no practical strategies prevent twin calves)
- any assistance required to deliver a calf (seek advice if >6%)
- any calf born dead or died within 24 hr of birth (seek advice if >1%)
- RFMs, that is membranes visible externally on the day of calving (seek advice if >4%)
- vaginal discharge or pus discharging from the vulva more than 14 days after calving (seek advice if >6%)
- lameness, or cows not bearing full weight on at least one leg which affects walking (seek advice if >3% of first calvers or >2% of older cows)
- abortions (seek advice if >5%)
- other health problems, including ketosis, displaced abomasum or cystic ovaries (seek advice if >5%).

Cow lameness can reduce intakes, milk yield and hence profits.

Some infections that cause abortions in cattle can also infect humans, so only handle aborted foetuses and membranes with disposable gloves and avoid contact with vaginal discharges from aborted cows. Bury the foetus and membranes ensuring that dogs are not allowed access to them.

7.5 Lameness

Foot problems can be caused by the environment (continuously wet floors, uneven or broken cement), poor sanitation, infectious organisms or nutritional imbalances. With tethered cows, annual trimming can extend herd life by at least a year.

Overgrowing will cause the toe angle to decline and foot length to increase, placing more weight on the heels, thus creating discomfort and walking difficulty. The rear feet appear to be more subject to disorders. High levels of concentrate feeding can lead to a condition called laminitis, which causes softening of the tissue between the claws of the feet. Trauma injuries can occur on uneven concrete floors, leading to swollen hocks, knees or even hips.

Inside the hard, outer layer of the hoof wall and sole, there is a sensitive layer rich in blood vessels and nerves. If a cow stands on a stone, or some other small hard object, its sole bends upwards over the stone, severely squeezing the sensitive layer. This can cause bleeding within the claw, and subsequently pressure, pain and lameness. Bruising is identified in a well-cleaned sole as pink or dark red flecks. Very soft feet, due to moist or wet conditions, are more prone to bruising. If the sole is thin due to excessive wear, it offers less protection to such damage. Such wear can result from standing on very rough concrete floors or animals being bullied by dominant animals or even excessive turning on floors when on heat.

Bruising will repair with time, but rest is important. Particularly severe bruising may need some form of relief from the pressure of body weight and walking. If bruising is largely confined to one claw, glue-on plastic or leather lace-up shoes can be fitted. Preventative measures include ensuring floors are not too abrasive, with all stones and broken pieces of concrete removed. Hoof trimming will also assist. Foot baths containing formalin (for hardening hooves) or sprays of zinc sulphate solution (for treating sore feet) are also useful.

Providing cows with soft bedding, such as a dirt lounging area or rubber mats improves cows' comfort and reduces feet and leg problems. Other factors influencing sore feet resulting from acidosis are discussed in Chapter 6.

7.5.1 Assessing cow lameness

Lameness is an increasing problem in both grazing and housed cows, which has economic implications. Locomotion scoring from 1 to 5 (for increasing lameness) is a new tool (Sprecher *et al.* 1997), which provides a quick measure of the cow's ability to walk normally. These scores are presented in Table 7.1. Observations should be made of cows' standing and walking (gait), with emphasis on their back posture, and should be made on a flat surface that provides good footing for cows.

Locomotion scores of individual cows can be used to select cows for hoof examination before they become clinically lame. Those with scores of 2 and 3 are

Table 7.1. Locomotion score guide based on observations of back posture and behaviour when walking

Score	Clinical description	Back posture	Assessment
1	Normal	Flat	Cow stands and walks with a level back. Gait is normal.
2	Mildly lame	Flat or arch	Cow stands with level back, but arches when walks. Gait is slightly abnormal.
3	Moderately lame	Arch	Stands and walks with arched back. Short strides with one or more legs.
4	Lame	Arch	Arched back is always evident and gait is one deliberate step at a time. Cow favours one or more legs/feet but can still bear some weight on them.
5	Severely lame	Three–legged	Cow demonstrates an inability or extreme reluctance to bear weight on one or more legs/feet.

(Source: Sprecher *et al.* 1997)

considered sub-clinically lame and their hoofs should be examined and trimmed to prevent more serious problems. Scores of 4 and 5 represent those cows that are clinically lame. The higher the lameness score, the greater the reduction in feed intake and milk yield and the poorer the body condition. For example, a score of 4 can reduce DM intakes by 7% and milk yields by 17%, while a score of 5 can reduce DM intakes by 16% and milk yields by 36%. Advice should be sought if more than 3% of first calving cows, or more than 2% of older cows, show signs of lameness.

7.6 Mastitis

7.6.1 Causes of mastitis

Mastitis is an inflammation of the udder caused by a variety of microbes, mostly bacteria, that gain access to the interior of the mammary gland through the teat canal. These microbes live on the cow, its udder and in its environment, including the floor, faeces, soil, feedstuffs, water, plants, and milking equipment and utensils. In response to these bacterial invasions, cells move from the blood stream into milk in order to fight the infection. Fortunately, these organisms are normally killed by pasteurisation and thus seldom cause disease in humans, unless the equipment is faulty or if raw milk becomes contaminated with these organisms.

There are over 100 types of microbes that can cause mastitis and, for convenience, they are grouped into two main types:

1. **Contagious bacteria.** These spread from infected quarters to other quarters, such as *Streptococcus agalactiae* and *Staphylococcus aureus.*
2. **Environmental bacteria.** These are commonly present in the cow's environment, such as environmental *Streptococcus uberis, Streptococcus galactidae* and the coliforms *Escherichia coli* and *Klebsiella.*

At low levels of infection, the disease may go unnoticed in the form of sub-clinical mastitis. However, it eventually becomes sufficiently severe to be classed as clinical mastitis, leading to pathological changes in the mammary tissue and physical, chemical and bacteriological changes in the milk. If the infection is not cleared up, chronic mastitis may result. As long as it persists, infected quarters can lose up to 25% of their potential milk production and produce only poor-quality milk.

The economic importance of mastitis is very high, through reduced milk production and treatment costs. Losses caused by clinical mastitis include discarded milk with antibiotic residues, drug and veterinary costs, the sale or possible death of infected animals, udder damage and the interruption to breeding improvement programs.

Clinical mastitis is easily recognised through visible abnormalities in the milk and udder. The sub-clinical form cannot be detected visually and requires tests to identify the microbes or their products of udder inflammation, such as somatic cells in the milk.

For each case of clinical mastitis in the herd, there will be 15–40 sub-clinical cases, while most clinical cases are preceded by infection at the sub-clinical level. Sub-clinical mastitis tends to cause a herd problem, is of long duration, reduces milk production and adversely affects milk quality. More importantly, it constitutes a reservoir of mastitis organisms that may spread to other cows in the herd.

Mastitis may be attributed to poor management, improper milking procedures, faulty milking equipment, inadequate housing and breeding for ever-increasing milk yield. Climate, season, herd size, type of housing, nutrition and stress all influence the incidence of mastitis. These all interact with genetic and physiological factors such as the stage of lactation, milk yield, milk flow rate and pregnancy.

7.6.2 The importance of the teat

The teat canal is the natural barrier to infection, but it is not always able to resist invasion by bacteria. The condition of the milking machine and how it is used, and damage to the teat canal orifice all influence the rate of mastitis infection.

The physical condition of the cows' teats is an excellent indicator of the state of the environment, the milking management and the milking system used on a dairy herd, and can also be used as an indicator for the risk of intramammary infections. The risk of mastitis is a 'numbers game' – greater numbers of bacteria near the teat end increase the risk of infections occurring. Teat sores and cracks provide sites where bacteria can multiply. They can be painful to the cow, causing her to kick and mess more frequently during milking, and to have poor milk let-down.

7.6.3 Environmental mastitis

Most cases of environmental mastitis occur within a few weeks of calving, when the cows' natural defence mechanisms are low and their udders have been in contact with mud and manure during calving. However, exposure of teat ends to environmental bacteria can occur at any time: before heifers have even calved, during calving, at milking time or in paddocks or yards during the lactation or dry periods. During lactation, factors that predispose cows to infection with environmental bacteria include milking udders that are wet or dirty, or administering intramammary infusions if the teat orifice is not sterile. During the early and late dry periods, the absence of the keratin plug in the teat canal may make cows highly susceptible to infection.

Control is best achieved by deceasing exposure of the teats to these organisms, through keeping cows in a clean environment. With *Streptococcus uberis*, which commonly occurs during the dry period, antibiotic treatment is not always efficient and chronic cases must be culled.

Good udder hygiene, correct use of good milking machines, dipping teats after milking and dry cow therapy are the keys to controlling environmental mastitis. In addition, heifers should be milked first, followed by uninfected cows with infected cows milked last. The steps outlined in Table 7.3 provide a comprehensive plan for mastitis control.

7.6.4 Contagious mastitis

The main mechanism of transmission of contagious mastitis is spread of pathogens from cow to cow at milking. These bacteria live on the teat skin or in the udder. Spread occurs when infected milk contaminates the teat skin of clean quarters or other cows. This can be by milk on milkers' hands or teat cup liners, through splashes or aerosols of milk during stripping, and by cross flow of milk between teat cups.

The spread of mastitis infections can be minimised by good hygiene, keeping teat ends healthy, using milking equipment that is operating well, and disinfecting teat skin after milking.

7.6.5 The dry period

The method used to dry off cows can influence how many udder infections establish during the dry period. The aim is to shut down milk secretion and seal the teat canal as rapidly as possible – this usually takes about 2 weeks. Most new infections occur in quarters where the teat canal has not sealed.

Drying off should be done abruptly. Cows should be milked as usual up till drying off. Once-a-day milking or milking every second day will greatly increase the risk of mastitis infection. The risk is higher because the stimulation provided by milking the cow impedes the ability of the teat to seal the canal.

Cows should also be removed from the sights and sounds associated with milking. These all stimulate the cow to let milk down, which leads to the breaking of the teat seal, which is vitally important in preventing infections during the dry period.

Special care needs to be exercised when using treatments at drying off. Dry cow therapy (DCT) when incorrectly administered can introduce bacteria into the teat.

After drying off the udders, the cows need to be checked each day for a week. Swollen quarters could indicate infections, which need to be treated so that they do not persist and create additional problems in the future.

7.6.6 California mastitis test (CMT)

This is a test performed on milk samples to indicate the level of somatic cells in the milk. It is sometimes referred to as the rapid mastitis test. It involves milking each suspect cow into a plastic container with four wells each containing a reagent: a detergent that bursts the somatic cells (from mastitis) in the milk. Equal quantities of a reagent and the milk sample to be tested are mixed and stirred for approximately 10–15 seconds. Any thickening or development of floccules in the solution indicates excessive somatic cells

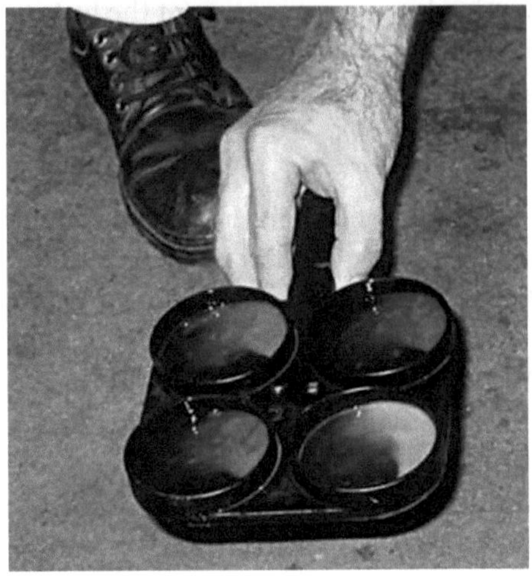

The rapid mastitis test is the basis of good mastitis management programs (Malaysia).

present in the sample. The greater the reaction the greater the somatic cell count (SCC), which increases three times for each unit increase in score (Table 7.2). The test is subjective, but is an excellent indicator of SCC, particularly when used as a cow-side test to detect problems in individual quarters.

7.6.7 The role of mastitis therapy

Despite good preventative procedures, new infections will occur while spontaneous recovery may also occur in some cases. However, drug therapy is the main treatment, with culling of chronic cases that resist antibiotics. Successful drug therapy requires the drugs to reach all sites of infection in the affected quarters, to remain at adequate therapeutic levels for sufficient time and to kill all the infecting microbes.

Table 7.2. Reaction scores, average somatic cell counts (SCC) and their description when read at 10 seconds in a California mastitis test

Score	Meaning	SCC (000 cells/mL)	Description of visible reaction
0	Negative	100	Mixture remains liquid and homogenous, with no evidence of thickening
T	Trace	300	Slight thickening when paddles tipped backwards and forwards. Disappears over time.
1	Weak positive	900	Distinct thickening, but no gel formation. May disappear after 20 seconds.
2	Distinct positive	2700	Mixture thickens immediately with gel moving towards centre when paddles shaken. Mixture settles on bottom of well when rotation stopped.
3	Strong positive	8100	Gel is formed with centre of surface elevated above main mass, even after rotation is stopped

7.6.8 Lactation therapy

The most effective treatment protocol will depend on previous experiences of the veterinarian, clinical signs and environmental management. Drugs should be chosen according to diagnoses of infective organisms, while the CMT allows monitoring of sub-clinical cases. For acutely ill cows, treatment with oxytocin and frequent milking out will aid recovery. Combination therapy, with both intramammary and systemic drugs, can improve cure rates of chronic cases.

7.6.9 Dry cow therapy

This involves long-acting drugs given to all cows, whether infected or not. There is no evidence of increased resistance to drug treatment through the use of dry cow therapy. Every udder of all cows must be treated, while reducing energy intake to ensure cows cease lactation. Prior to treatment, the teat must be disinfected with 70% alcohol to reduce the likelihood of introducing microbes with the dry cow therapy. The administration and drug withdrawal instructions should be followed carefully.

It is false economy not to treat infected lactating cows, and all cows at drying off. The extra milk produced is worth considerably more than the reduced milk yields of infected cows.

7.6.10 Preventing mastitis

As already stressed, the cost of doing nothing is very high, so prevention is better than cure. There are many procedures that should be routinely followed to reduce the high economic impact of mastitis.

Suggested trigger points for seeking professional assistance are:

- six clinical cases of mastitis for every 100 cows during calving
- five clinical cases of mastitis for every 100 cows in the first month after calving
- two clinical cases of mastitis for every 100 cows per month for the remainder of the lactation.

7.6.11 Recommended milking procedures

The 15 min each day when cows are milked is a prime time for increased susceptibility to new mastitis infections. Correct milking procedures are important, regardless of whether cows are milked by hand or machines. Preparation of teats and udders for milking reduces the number of contaminating microbes on teat skin, and stimulates milk let-down. As well as reducing this contamination, it decreases residual milk left in the udder at the end of milking, increases milk yield, decreases milking time and reduces the spread of contagious and environmental organisms that cause mastitis and reduce milk quality. Key milking procedures include the following:

- Provide cows with a clean, stress-free environment. The cows should not be frightened or excited before milking because this stress will release hormones that interfere with normal milk let-down and reduce the cow's resistance to mastitis.

- When washing the cow or cooling it down, ensure that at least 30 min elapses before milk harvesting, to minimise the likelihood of dirty water dripping off the teats during milking
- Check the udder and foremilk for mastitis. Clinical mastitis can be visually detected from clotty, stringy or watery milk or hot, hard and enlarged quarters.
- Wash the teats and lower surfaces of the udder. Correct washing and massaging releases hormones for milk let-down. A sanitising solution should be used in the bucket with an individual cloth or paper towel to wash the udder. Do **not** use a common cloth because it allows for cross contamination between cows.
- Use a pre-milking teat dip (optional). Disinfecting teats before milking reduces the rate of infection of environmental mastitis by 50%, as well as reducing contagious mastitis. The recommended procedure is to wait 20–30 seconds after pre-dipping teats, then thoroughly dry the teats prior to applying teat cups.
- Dry the teat thoroughly. Using excess water to wash teats and udders and not drying them properly results in water laden with microbes draining down into the milk bucket (with hand milking) or being drawn into the teat cups of milking machines.
- Hands should never come into direct contact with milk.
- Attach teat cups within 1 min. Attachment must be done carefully to prevent entrance of excessive air into the milking system. Maximum internal udder pressure is reached within 1 min after udder preparation and lasts for about 5 min. Because the majority of cows will milk out in 3–5 min, the 1 min delay in attaching the milk cups maximises the effectiveness of the milk let-down hormone oxytocin.
- Observe milking units closely while attached to ensure that they are adjusted correctly and to aid in preventing cup slip. Slipping or squawking cups slow down the milking operation and may cause machine-induced infections. Liner slip can allow tiny droplets of milk to be quickly propelled against the end of the teat, which may penetrate the teat canal. Because most liner slip occurs towards the end of the milking, the chance of these organisms being flushed out is reduced, leading to an infected quarter.
- Avoid irregular vacuum drops due to sudden air intakes when connecting or removing the milk clusters.
- Shut off the vacuum before removing the teat cups at the end of milking. Never pull the milking unit off the udder while still under vacuum because this causes the same problem as squawking cups, leading to infections (referred to above). A minute or two of over milking with a properly functioning milking unit does not pose a major risk in term of mastitis. If one quarter milks out ahead of the others and if the teat cup remains attached to the teat without slipping, it should be left on because careless removal may permit air to enter, possibly leading to liner slip-induced infection. If there is a probability of cup slip on a particular teat, then the cluster should be lifted gently to seal the short milk tube of the liner over the ferrule of the claw to shut off the vacuum before detaching it from the teat. Never remove a teat cup while under vacuum.
- Don't over milk any cow; 8 min is maximum milking time, even for a slow milker.
- Dip teats with an effective teat dip. This is the most important way to reduce new infections of environmental mastitis. The goal should be to dip the entire surface of the teat that comes into contact with the teat cup liner. Teat spraying never covers the entire teat surface. Teat cups should be kept clean and sanitary, but teat dip should

never be poured back into the original container because the strength of the used teat dip may be reduced and allow organisms to survive.

- Disinfect teat cups between cows (optional). Teat cups can spread organisms from one cow to another. If they are to be disinfected between cows, two teat cups should be dipped at a time in a disinfectant solution. The solution should be changed when it becomes cloudy to ensure mastitis organisms are killed rather than spread.
- When the milking session is complete, clean and disinfect the teat cup liners properly, using hot water plus detergent followed by hot water plus sterilising agents.
- Allow each cluster to drain between milking sessions.
- If using a bucket milking machine, ensure the bucket is hung upside down so it can drain properly before reuse.
- Feed the cows immediately after milking to ensure they do not lie down for at least 30 min.
- Ensure the cows lie down on clean bedding or a clean floor.

When milking mastitis-infected cows:

- Draft out clinically affected cows and milk them last.
- Use gloves when milking mastitis-infected cows.
- Rinse and sanitise the milk cluster after milking each mastitis-infected cow and don't touch any other cluster or cow for at least 20 seconds.

To reduce exposure to environmental mastitis:

- Calve cows down in a clean and dry area.
- Be alert to the number of cases of mastitis occurring in freshly calved heifers; this is an indicator of the state of the calving area.
- Bring cows into the dairy for milking out and checking, certainly within 24 hr post-calving.
- Look for swollen quarters and check them for heat and pain.
- Check milk from all quarters of freshly calved cows.
- Take special care of induced cows; that is, cows that have been treated to calve down early.
- Take special care with pre-milking preparation; that is, teat cleanliness and ensuring all milking equipment has been sterilised and allowed to drain.

7.6.12 Recommendations for milking machines

Mastitis can result from improper design, malfunctioning, misuse or improper cleaning of milking machines. If they are properly designed, maintained and used, milking machines will account for only 6% of all udder infections.

Although they differ in design and size, all milking systems have the same basic components. These include systems to produce a vacuum, pulsation, milk removal and milk handling, which are all run by an electrical system. With respect to mastitis control, it is the milk removal system – in particular the teat cup liner slips – that are associated with milking machine-induced infections. A problem with liner slip exists if there are more than 5–10 liner slips per 100 cows milked. Some of the causes of this include poor cluster alignment, poor liner design, uneven weight distribution clusters, blocked air vents at the claw and flooded milk lines. In addition, cup slip results from inappropriate liner design, cluster weight, vacuum levels, vacuum fluctuations, milking wet teats, absence of hose support arms, over milking and teat size.

The steps in milking machine maintenance include the following:

- Ensure the milking machine technician closely follows the maintenance schedule for testing and servicing the machine.
- Regularly check the pulsation rate of the machine: 60 pulsations/min is the recommendation.
- Regularly check the vacuum pressure and ensure it is not too variable during milking.
- Ensure the diameter of the vacuum line corresponds with the capacity of the vacuum pump.
- Ensure the vacuum line is not too long; that is, the distance between the vacuum pump and the teat cups.
- Routinely replace any cracked milk tubes.
- Replace the milk cluster rubber liners after 2500 milkings, using the following simple formula:

$$\text{Replacement age (days)} = \frac{2500 \times \text{number of claws}}{\text{herd size} \times \text{milkings/day}}$$

Therefore, for a 40 cow farm milking twice daily, with one mobile milking machine with two claws, the rubberware should be replaced every (2500 × 2)/(40 × 2) or every 62 days.

The following treatment protocol for mastitis should be followed by the veterinarian or dairy cooperative animal health staff:

- Regularly monitor each cow for sub-clinical mastitis, preferably using the CMT.
- Apply antibiotics under veterinary supervision, ensuring it is the most appropriate for the mastitis infection.
- Use the full course of antibiotics, as specified on the label.
- Clearly mark treated cows.
- Keep records of treatment to clinically affected quarters.
- Ensure the withholding periods (for drug use) are followed before mixing the milk with that of the rest of the milking herd.
- Discard milk from all quarters of cows that receive treatment.
- If a cow fails to respond to full treatment, consider a repeat course of the same or another antibiotic, drying off the infected quarter, drying off the cow or even culling the cow.

7.6.13 Controlling mastitis

Mastitis is best viewed as a herd problem rather than an individual cow problem. Unlike other cattle diseases, such as brucellosis or tuberculosis, mastitis cannot possibly be eradicated on a large scale. Each individual herd is the unit of control by itself. The level of mastitis in a herd has nothing to do with the level in a neighbouring herd.

Table 7.3. Comprehensive mastitis control

Management task	Specific actions
Milking hygiene	Milk teats that are both clean and dry
Milking machines	Stable milking vacuum No slipping or squawking of liners Shutting off vacuum before removing
Post-milking teat dipping	Immediately after removing teat cups Full teat immersion, not spraying
Drying off	All quarters of all cows after last milking
Treatment of clinical cases	Early detection and treatment Maintain records
Culling	Cull chronic cases
Environment	Clean and dry Uncrowded and well ventilated
Herd replacement	Test new animals before adding to herd Check new animals regularly

(Source: Thirapatskun 1999)

A control program for each herd must be practical, easy to understand, highly effective, increase economic returns, reduce new infections, shorten the duration of pre-existing infections, provide tangible evidence that clinical mastitis is reduced and be subjected to easy modifications as improved methods are developed.

The level of infection in a herd must be known to be able to assess its seriousness, and the source and risk of spread of infections. Prevalence of mastitis indicates a level of infection or the proportion of cows or quarters infected with the disease at a given time. Regular monitoring of each farm is necessary. If the rate of infection is reduced, the level of infection will fall, though very slowly. If the duration of infection is effectively shortened, the level of infection will soon be reduced, provided that no new infections occur. To keep mastitis at a low level, it is necessary to prevent as many new infections as possible and then shorten the duration of those that do occur and finally eliminate the existing infection. Table 7.3 summarises an effective approach to controlling mastitis.

7.6.14 Heifers

Udder infections among heifers are quite common, even in very young animals and can cause inflammation and tissue damage prior to calving. Sources of infection can be insects, sucking among calves (particularly those fed mastitic milk), and microbes in the mouth and on the skin and coat. Strategically treating pregnant heifers using drug therapy is effective. Extreme caution must be exercised to make sure the teat ends are sanitised before injecting antibiotics.

7.6.15 Establishing goals

As with any control program, it is important to establish realistic and achievable goals. These can be based on milk somatic cell counts, herd milk bacteria counts, proportion of cows with sub-clinical or clinical infections and the volume of discarded milk.

A high incidence of clinical mastitis usually indicates a major deficiency in management, such as an unsanitary milking area, inadequate milking hygiene, faulty milking machines or purchases of infected cows. It is important to determine whether the mastitis is environmental or contagious, so that the most appropriate measures can be taken.

Evidence of progress will often require several months, so regular monitoring is necessary to maintain motivation and identify areas where the program may have to be re-evaluated. As with any successful control program, follow up visits are necessary. Bulk milk somatic cell counts and inflammation and infection percentages on a cow and quarter basis are useful tools.

8

Reproductive management

This chapter explains the impact of both nutritional and non-nutritional factors on the reproductive performance of the dairy herd.

The main points in this chapter

- Understanding the fertility timetable for the milking cow assists with managing reproductive procedures.
- 100 day-in-calf rate and 200 day-not-in-calf rate are two good measures of reproductive performance. Submission rate and conception rate are the other two important measures.
- Nutrition is just one of many factors affecting reproduction. These factors can be categorised into their degree of difficulty for the farmer to influence.
- The key nutrition factors are feed intake, body condition and heifer live weight.
- Heat detection, length of the voluntary waiting period and artificial insemination practices are three other factors.
- Farmers must set priorities to tackle fertility issues, dealing with the most important first. It is possible for a healthy, high-yielding cow to require up to eight inseminations to conceive.
- The simplistic answer for any problem cow is 'not enough insemination', but when is enough enough?

Improved reproductive performance provides many benefits to farmers such as:

- higher average milk yields each day – cows with poor reproductive performance will spend more of their time in late lactation, when daily milk yields are lower
- fewer cows that have become excessively fat because they have failed to conceive
- less compulsory culling of cows failing to become pregnant
- fewer cows with long dry periods
- reduced insemination and semen costs
- heifers calving at a younger age
- increased number of calves produced each year, thus providing more animals for sale or as replacements for the milking herd
- more efficient feed utilisation as a result of the above benefits
- more profits, less work and less worry.

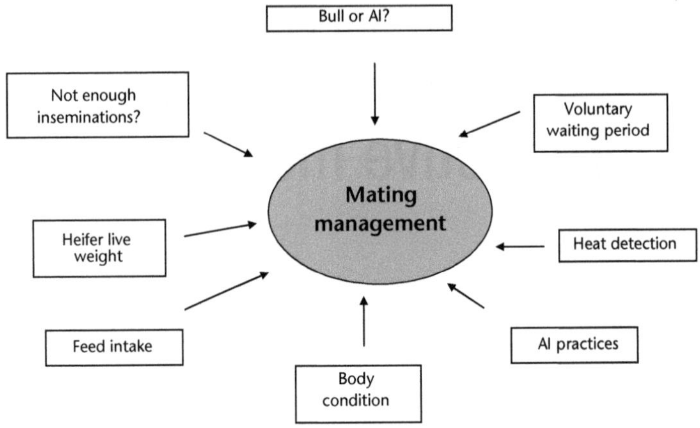

Figure 8.1. The key factors influencing the milking herd's reproductive management

The Australian dairy industry has a structured program to maximise the rate at which cows get in calf, called InCalf (Morton *et al.* 2003). It is a framework for a fertility management plan to make sure all the day-to-day actions work towards a goal of better reproductive performance. The 'fertility for life' cycle involves calf and heifer rearing, first mating, pregnancy and calving. Then for subsequent lactations, it involves matings and pregnancies, calvings and, eventually, culling. Success will require attention throughout the cycle. Basically, any management plan should answer an important question: today has everything been done to ensure high reproductive performance? From the day a calf is born, farmers control the factors influencing her future fertility, hence whether she gets in calf on time, every time.

There is no simple recipe because every farm is different and will require its own plan. InCalf focuses on the key drivers of:

- calf and heifer rearing
- body condition and nutrition
- voluntary waiting periods of no more than 50 days
- heat detection
- artificial insemination (AI) technique and sire selection
- bull management
- cow health
- heat stress management.

8.1 Measures of reproductive performance

8.1.1 Collecting permanent records on each cow

It is important to develop a system for collecting permanent records of each adult cow in the herd. This facilitates extracting data to calculate the above measures. Permanent records should include the following information:

- cow identification number, which should also be on an ear tag (preferably two tags, in case one falls off)
- birth date
- ancestry, such as dam and sire number and breed type, maybe grand dam and grand sires number and breed type, and cow's breed type
- vaccination dates
- important disease problems during rearing
- for each lactation:
  - lactation number
  - dry off date of previous lactation
  - calving date
  - each insemination date – this is generally the important data that is not routinely recorded
  - the date when confirmed pregnant by rectal palpation
  - the estimated age of the foetus
  - any estimates of live weight or body condition
  - peak milk yield or estimated full lactation yield
  - any diagnosed health problem.

8.1.2 The fertility timetable for the milking cow

The fertility cycle can be best understood by following the recommended reproductive timetable:

- calving, with minimum difficulty
- involution of uterus takes 21 days
- follicular development commences 14–21 days after calving, in a well-managed cow
- voluntary waiting period, or days after calving with no insemination, should not exceed 50 days
- cycling occurs every 18–24 days
- first insemination should be 50–80 days post-calving
- pregnancy takes 282 days
- dry off cow 50–60 days pre-calving
- transition period to calving for 14–21 days.

The annual fertility cycle incorporates replacing 20–30% of the herd with heifers each year and culling up to 10% of the cows for poor reproductive performance, and this includes 5% of the herd that will never get pregnant, even though they are healthy.

Before discussing the effects of herd and feeding management on reproductive performance, it is important to quantify reproductive performance.

8.1.3 Introducing 100 day-in-calf and 200 day-not-in-calf rates

There is an old saying 'If you can't measure it you can't manage it'. Benchmarks make it is possible to respond when measures indicate reproductive performance can be improved and to assess whether it has improved as a result of management changes. The best

Routine pregnancy diagnosis is an integral part of good reproductive management (Vietnam).

measures describe the rate at which cows get pregnant when mated and the number that remain non-pregnant following mating.

For year-round calving herds, the two useful measures of reproductive performance are:

- **100 day-in-calf rate.** This calculates the percentage of the cows in the herd that become pregnant by 100 days after calving. It also describes how many cows will calve within about 13 months of their previous calving.
- **200 day-not-in-calf rate.** This calculates the percentage of cows not pregnant by 200 days after calving. Farmers want as many cows as possible to calve no more than 15–16 months after their previous calving.

To calculate these two measures of reproductive performance, one must know accurately when cows become pregnant. It is best to base this on pregnancy testing when cows are between 5 and 15 weeks pregnant. To achieve this, a herd pregnancy testing routine is necessary. Cows not returning to service should be examined on the next pregnancy testing day until they are confirmed pregnant. Pregnancy testing cows every 1–2 months allows most accurate monitoring.

Two other measures are frequently used to describe reproductive performance. These are average calving to conception interval (or days open) and inter-calving interval. These measures are not ideal because there are three major problems with their use, namely:

- They are only averages and don't indicate how many cows actually had long calving to conception intervals. Such problem cows require attention.
- They do not take into account cows that do not become pregnant at all. Obviously these non-pregnant cows are also problem cows.
- These measures can include cows that are routinely inseminated many months after calving, such as high-producing cows, which are normally culled as non-fertile.

8.1.4 Submission rates

The submission rate (SR) is the percentage of the herd which received at least one insemination within a specified number of days after calving.

To achieve a high 100 day-in-calf rate, a high percentage of cows in the herd must be submitted to insemination without undue delay after calving. The 80-day submission rate is the percentage of cows that receive at least one insemination by 80 days after calving.

8.1.5 Conception rates

The conception rate (CR) is the percentage of inseminations that are successful and result in pregnancy. To determine this value, dates of every insemination or natural matings to a bull must be collected and documented. The date of conception for each pregnant cow must also be recorded, based on pregnancy diagnoses, and this will be the date of the last insemination. The conception rate is then the number of conceptions or natural matings that resulted in a successful pregnancy (more than 40 days previously) expressed as a percentage of the total number of inseminations or natural matings for those cows. Conception rate to first inseminations is the number of cows conceiving at the first insemination expressed as percentage of the total number of cows included in the calculations. Herds can have high conception rates but poor 100 day-in-calf and high 200 day-not-in-calf rates.

The conception rate is only one aspect of reproduction because herd performance is also dramatically affected by:

- the time from calving to first insemination
- the percentage of returns to service that are detected.

A cow breeding wheel to monitor when cows are successfully bred (Vietnam).

Table 8.1. Achievable measures of reproductive performance in year-round calving herds

Measures of reproductive performance	Seek help	Top farmers achieve
100 day-in-calf rate	<45%	58%
200 day-not in-calf rate	>19%	13%
80-day submission rate	<61%	73%
Conception rate	<43%	51%

(Source: Morton *et al.* 2003)

High levels of reproductive performance can be achieved with only moderate conception rates when submission rates are high and a high percentage of returns to service is detected. High submission rates are essential for high 100 day-in-calf rates.

Data from Australian dairy farm surveys can provide a guide for actual and achievable reproductive performance in year-round calving herds. These are presented in Table 8.1.

These large-scale surveys identified six factors that have large influences on herd reproductive performance. Three are non-nutritional and three are nutritional. They are:

- **The length of the voluntary waiting period.** This is the number of days delay after calving before farmers begin inseminations. This is 50–55 days in the herds with the best fertility.
- **Mistakes in heat detection.** Farmers can make two types of mistakes: they can diagnose heat in cows not on heat (called a false positive) or miss a heat identification (undetected heat). Missed heats are more common. The higher the heat detection rate, the higher the submission rate. Farmers with over 80% heat detection rates had 73% 80-day submission rates.
- **AI practices.** There are many skills in AI, but discussion of these is outside the scope of this manual. Good first insemination rates were 45–48%.
- **Body condition.** Cows calving at condition scores of 4.5–5.5 (where 1 = emaciated and 8 = extremely fat) had higher 100 day-in-calf rates (54%) than those calving at less than 4.5 (41%). Cows calving in very high score (6.0 or more) may lose condition more rapidly after calving and can suffer reduced fertility.
- **Feed intake.** Better fed cows have higher fertility. Better feeding can improve 100 day-in-calf rate from 41% to 57% and reduce 200 day-not in-calf rate from 15% to 9%.
- **Heifer live weight.** The occurrence of the first oestrus in yearlings depends on live weight. So better feeding practices in early life will lead to younger age at first calving in virgin heifers. These heavier animals will also cycle earlier after calving.

8.2 Non-nutritional factors affecting reproduction

Reproductive performance is affected by many factors. Nutrition is only one possible cause of poor reproductive performance. In some herds, nutrition is not the most

Table 8.2. Factors affecting submission and conception rates with AI (or natural mating) in tropical small holder herds. Italics refer to natural mating.

Ease of influencing	Submission rate (Cows being submitted for insemination or mating)	Conception rate (Services per conception)
Difficult Outside farmer control	Mineral deficiencies (defining what minerals)	Semen quality Infectious diseases (requiring government support) Mineral deficiencies (defining what minerals)
Possible Requires some management skills and/or capital	Heat stress Genotype × environment interactions: Feeding to cows potential Genetic tendencies for lactation anoestrus Support of infrastructure, e.g. facilities	Semen handling AI technique Heat stress
Easy Requires little management skill and/or capital	Deferral of insemination or voluntary waiting period Heat detection Age Heifer size (and age) at first calving *Bull performance* *Bull:cow ratio* Body condition at calving Body condition loss after calving Induced calvings Post-calving nutrition (energy, protein, fibre) Health problems: Uterine infections Lameness Retained foetal membranes Milk fever Cystic ovaries	Timing of insemination *Bull fertility* Body condition at calving Body condition loss after calving Post-calving nutrition (energy, protein, fibre) Diseases affecting fertility: Uterine infections Other diseases

(Source: Moran 2005)

important cause of poor performance. Some of the other causes can be divided into those that affect submission rate and those that affect conception rate.

The many factors affecting submission and conception rates in tropical SHD herds are listed in Table 8.2 and categorised into:

- difficult to influence (generally outside farmer's control)
- possible to influence (require some management skills and/or capital)
- easy to influence (require little management skill and/or capital).

It is easier to improve submission rates than conception rates. For example, submission rate may be increased from 40% to 80% with careful management, whereas conception rates may only improve from 30% to 45%. Many of those factors influencing submission, but not conception, rates are behavioural, such as decreasing the length of voluntary waiting period and making more effort to improve heat detection.

8.2.1 Heat detection in tropical herds only using AI

Observations of oestrus are more difficult in the tropics due to anoestrus resulting from poor nutrition and/or intensive suckling. Furthermore, the oestrus period is shorter (10–12 hr), signs are less pronounced or mainly shown at night (in buffaloes or local cattle) when farmers are less keen on, or active in, heat detection.

For practical purposes, most small holders use AI rather than natural mating. This introduces another set of factors limiting herd fertility. Cows tied up in a stall can hardly express the most easily recognisable signs of oestrus, such as mounting and being mounted.

Cows show signs of heat when:

- they are 18–24 days after their last heat (if they are still non-pregnant)
- they stand to be mounted
- they attempt to mount other cows
- they are restless and bellow
- their feed intake is reduced
- they have poor milk let-down
- stringy mucus is seen exuding from their vulva
- their vulva is red and swollen.

The average duration of heat is about 14 hr in normal weather conditions. Heats can be as short as 2 hr and as long as 28 hr. Twice daily observations are then essential to catch short heats. Observations in the cool of early morning are more likely to detect heat than those in the middle of the day.

The best conception rates occur following insemination 4–12 hr after the first signs of heat are observed. However, the problem is knowing at what stage of oestrus the particular heat was first detected.

Heat detection can be improved with:

- routine night observations
- interpreting cow behaviour
- checking records for days since previous heat (for closer observation)
- using heat detection aids in larger herds, although tail paint is a cheap effective aid for most farmers
- using oestrus synchronisation as a management aid.

SHD farmers who continually tether their cows are disadvantaged because the cows cannot move around to show the most easy to interpret behavioural signs of heat. Having a small yard in which to let the cows out, say before milking, will aid detection. Using heat aids, such as tail paint or heat mount detectors, can improve detection rates. Vasectomised bulls or hormone-treated steers running with the mating herd are also useful in larger herds. Heat synchronisation can offer efficient use of labour as the work of heat detection and AI is shortened into planned, intensive periods.

Each month farmers need to identify cows that have calved more than 80 days before, but have not been detected on heat, and examine them. This is important if more than 60% of the herd are in this category. Some of these cows may have had an undetected heat, whereas others may not have been on heat and can be treated as non-cycling cows.

If most of these cows are in low body condition, their feeding management should be improved. Others may be suffering disorders such as cystic ovaries, infected uterus and lameness, thus requiring veterinary attention.

Other important factors should be taken into account, such as:

- body condition: too fat or too thin
- rapid loss in body condition
- small heifer size
- diagnosed health problems, such as cystic ovaries, uterine infections or lameness
- heat stress, as apparent from observed high respiration rates and/or high Temperature Humidity Index (see Chapter 12).

Assuming heat can be detected early, the message must be passed on and acted on by the AI technician. Most dairy cooperatives have a simple and effective system of communication between farmers and AI technicians. Assuming the technician can reach the cow to be inseminated, many of the factors listed above are relevant.

8.3 Nutritional influences on reproduction

8.3.1 Energy intakes and balance

Energy intake affects milk production, body condition and live weight. Numerous studies have investigated the effects of milk production, body condition score and live weight on fertility. Results have varied widely. To date, the effects of energy intake on fertility have not been fully clarified. Nevertheless, some effects are clear. The effects of energy intake can be separated into effects on submission rates and effects on conception rates.

Effects of energy intake on submission rates

Cows in poor body condition have delayed heat, hence poorer submission rates. At the other extreme, cows can be too fat at calving. Cows calving in too high condition scores are at greater risk of fatty liver syndrome, with associated reduction in appetite, excessive weight loss and increased likelihood of retained foetal membranes and/or non-cycling.

Extensive work is now focusing on the effects of nutrition in high-producing cows. Probably the best current explanation of the causes of delayed cycling among high-producing cows comes from North America. Under the systems of management there, it appears that condition loss after calving is more likely to cause delayed cycling than is the level of milk production in early lactation.

The problem is that milk production peaks before maximum feed intake, even where cows are fed *ad lib*. Therefore in early lactation, cows' energy requirements exceed their intakes. They are in 'negative energy balance' and so they lose body condition and weight. There is also a lot of variation between cows within herds. Higher producing cows will generally have deeper and more prolonged troughs than lower producing cows.

This negative energy balance is worst in the first few weeks after calving. Intake capacity steadily increases and cows eventually move into positive energy balance. When in positive energy balance, cows began to gain condition.

How does this impact on reproductive performance? Some research suggests that cows have their first ovulation about 10 days after the greatest negative energy balance. So if the negative energy balance is prolonged, first ovulation will be delayed and submission rates reduced. Greater negative energy balance may also cause longer delays to first ovulation.

Effects of energy intake on conception rates

Energy intake and condition score affect onset of cycling after calving. Early onset of cycling clearly increases submission rates. Furthermore, early onset of cycling increases conception rates.

New Zealand research suggests that conception rates are improved in cows that have had at least one heat before mating. This is regardless of the length of time between calving and the start of mating.

The ovary produces progesterone after a cow has ovulated. Cows inseminated on their second ovulation will usually have had much higher progesterone levels for the previous 12 days.

There may also be other explanations for the improvement in conception rates among cows having a heat before mating. Hormonal changes may explain the positive effect of a heat before mating on conception. It appears that high levels of the hormone progesterone in the 12 days before insemination improve conception rates.

Energy balance may also affect conception rates in other ways. Negative energy balance long before insemination reduces conception rates. In the ovary, egg development begins around 60–100 days before the heat where the egg is released. It may be that adverse conditions around that time can result in a defective egg being produced and released. If true, this means that severe weight loss in early lactation leads to poor-quality eggs when cows are inseminated. For late-calving cows, which are mated soon after calving, weight loss in the dry period could have similar effects.

8.4 Implications for herd and feeding management

8.4.1 A simplified system for feeding management

Cows in early lactation have high nutrient demands to produce milk as well as return to normal heat cycles. Because they have yet to reach full intake capacity until 2–3 months post-calving, many show high rates of loss in live weight and body condition, thus delaying their first heat.

In many Asian large herds (say with more than 20–30 milking cows), all milking cows are managed together as a single herd except for feeding their concentrate allocation, which is generally based on individual milk yield. In such herds, reproductive management would be easier if the milkers were split into two herds: the non-pregnant and pregnant. This would be essentially the same as grouping the cows into early-mid and mid-late lactation.

Pregnancy diagnosis, rather than non-return to heat, would be the most suitable method to decide on moving cows to the pregnant group. Early pregnancy diagnosis by skilled operators can identify cows that are 6 weeks pregnant, while it can also provide a good estimate of when cows conceived if tested at 15 weeks of pregnancy or less. Cows to be tested each month include those mated more than 6 weeks previously or those running with a bull for more than 80 days and not confirmed pregnant and cows previously diagnosed as pregnant but thought to have aborted since.

The non-pregnant cows should be offered the best quality forages and be more closely observed for signs of heat. Feeding management would be simplified if all cows in each group were offered the same amount of concentrates as well as forages. From our experiences with seasonal calving herds in southern Australia, there is little difference in the utilisation of concentrates for milk production whether all cows in early (or in late) lactation are fed the same amount of concentrates or fed depending on individual milk yields.

8.4.2 Management in late lactation

Good herd managers aim to dry off cows in the condition in which they want them to calve. How cows are fed in late lactation, and when they are dried off, will have huge effects on the condition score at drying off and at calving.

8.4.3 Dry cow management

Dry cows must not lose weight. Even if live weight does not seem to be declining, cows may still be losing condition. The increasing weight of the foetus and udder during pregnancy may mask losses of the cow's own body weight.

8.4.4 Management during the transition period

Decreases in appetite around calving have to be minimised. The cow's rumen needs time to adjust to the new diet. Digestion and intake can be reduced while the rumen is adjusting. This adjustment can start before calving.

In herds where milkers are fed moderate to large quantities of grains, lead feeding (gradual introduction of grain in the diet before calving) helps reduce the impact of dramatic change in diet. Lead feeding may also help retain condition in the last few weeks before calving. Anionic ('acid') salts may enable some forages to be fed to cows due to calve without increasing the risk of milk fever (see Chapter 6). Feeding some pasture may help reduce the impact of the dietary changes after calving.

8.4.5 Protein intakes

Excess protein

Some diets may lead to the protein content of total diet exceeding cows' requirements. The excess can reduce reproductive performance.

Most of the protein in forages is rumen degradable protein (RDP). Research work suggests that diets high in RDP depress conception rates. Imbalance between the amounts of fermentable carbohydrate and protein in the diet may be the key concern.

Diets with excess protein have effects other than on conception rates. It costs the cow energy to break down the excess protein: energy that could otherwise have been used for milk production and body condition. So, adverse effects of high protein diets on fertility may be due to an energy deficit.

Protein deficiency

The cow's highest requirement for protein is in early lactation. Requirements for both RDP and undegradable protein (UDP) increase after calving, especially among high-producing cows.

Many of the diets fed throughout Asia do not provide sufficient protein. This can also reduce reproductive performance.

8.4.6 Intakes of minerals

Calcium

There are suggestions that sub-clinical calcium deficiency in cows around calving can result in calving troubles, retained foetal membranes and inappetence. If correct, these effects could result in reduced reproductive performance.

If either milk fever or sub-clinical calcium deficiency is considered to be a problem, anionic salts could be used in conjunction with other dry cow management strategies. Cows due to calve should not be fed supplements containing sodium bicarbonate.

Magnesium

Supplementary magnesium fed to calving cows reduces the risk of grass tetany. It also appears to reduce the incidence of milk fever. Some advisers recommend this to avoid sub-clinical calcium and magnesium deficiencies in cows at calving.

Phosphorus

Phosphorus deficiency has been shown to cause delayed onset of heat in some studies but not others. Phosphorous-deficient cattle can also be energy-deficient in some circumstances (e.g. during prolonged periods of underfeeding). In these situations, effects of phosphorus deficiency cannot be readily determined.

In early lactation, phosphorus supplied in the many forages is insufficient to meet cows' requirements. Affected cows then use their bone reserves of phosphorus. This may be sufficient to prevent reproductive performance being affected.

Selenium

There are numerous reports of retained foetal membranes due to selenium deficiency.

8.4.7 Herd management

Achieving 100% success with AI

It is just not possible to achieve 100% of all cows getting back in calf no matter how many AI cycles a farmer undertakes. It is possible that a healthy, high-yielding cow may require up to eight inseminations to conceive. The difficulty for the farmer is deciding when enough is enough and when that cow should be culled.

Heat detection and insemination

Voluntary waiting periods should be no more than 50 days, after which cows should be inseminated when observed to be on heat. Although CR can be low (33% when 40 days post-calving, 45% when 60 days post-calving) cows can only conceive when inseminated. A voluntary waiting period of 80–90 days appears to be more the norm in most Asian countries.

There are two types of error in heat detection: missing a cow when on heat or inseminating a cow when not on heat. With the first error, SRs are low, while with the second error CRs are low. Small holder farmers (who depend on AI technicians to do the insemination) have an additional problem in that the technician must first agree that they also consider the cow to be on heat, before she gets inseminated.

8.5 Putting it all together – a plan for success

Reproduction is a complex issue and solutions to problems don't come in a single package, as many actions may be required to achieve targets. To ensure success, these actions need to be part of the overall farm management. It is easy to lose track of the length of time between when cows calve and when they should be first mated and how many times they have been mated this lactation.

8.5.1 Setting priorities to address fertility issues

- Put first things first, or what are the most important management practices required to achieve good reproductive performance?
- Small steps make can big gains.
- Attention to detail can make the difference.
- Solutions are often not expensive or time consuming.
- Keep accurate records, particularly of all inseminations.
- Have a plan for each stage of the cow's life cycle, because success requires a fertility life plan.
- Deal with the most important things first because some things are more important than others.
- Work on the things you can control before worrying about those you cannot control.
- Consider culling cows that have had many inseminations and are now producing little milk.
- Monitor performance carefully and set realistic targets (Table 8.1).

8.5.2 Some important decisions to make

- Select a suitable voluntary waiting period, ensuring it is short if milk yields decline rapidly in the herd or heat detection is poor.
- Assess the need to cull more severely by calculating the percentage of cows that have more than 16 months between calvings. If this is more than 15% of the herd, then some of the low-fertile cows need to be culled. Make a conscious decision on how long to continue mating any cow that is difficult to get in calf. This will depend on:
  - her current milk production
  - her age
  - her previous milking record
  - if she has clinical mastitis
  - if she has previously had difficulty getting back in calf
  - of most importance, the cost of keeping her compared with selling her and buying a replacement.

Heat detection must be optimised to limit missed heats, using both improved observation and greater use of heat records. In addition, feeding management must be addressed to ensure that:

- body condition scores are adequate at calving
- losses in body condition are minimised during early lactation
- target live weights are achieved for growing heifers.

The biggest factors affecting reproductive performance of individual cows are:

- problems at calving, such as abnormal calvings (assisted or twin calving) or heath problems (retained foetal membranes, vaginal discharges or lameness)
- cows having poor fertility in their previous lactation
- age extremes, in that very young (2 and 3 yr old) or old (7 yr or older) have poorer reproductive performance relative to middle-aged cows.
- generally speaking, a cow with a reproductive problem in one year is more likely to have a problem the following year.

One simplistic answer for a solution to any problem cow could easily be 'Not enough inseminations' but when is enough enough?

9

Genetics

This chapter discusses the role of genetics on small holder farm performance and profits.

The main points in this chapter

- The continued emphasis on temperate dairy genotypes, which are essentially exotic to tropical small holder farms, has had little impact on cow performance or farm profitability over the last few decades.
- The concept of genotype by environment interaction is introduced to explain the need to modify the farm environment to get the most benefit out of improved dairy genotypes.
- Several tropical dairy breeds have been developed, such as the Australian Milking Zebu and the Australian Friesian Sahiwal.
- The availability of artificial insemination has the greatest effect on the farmer's ability to initiate a breeding program.
- In Sri Lanka, despite the extensive use of AI, over 65% of the progeny are born from local bulls. This is attributed to the poor infrastructure of AI services as well as the lack of farmer confidence that AI will result in a pregnancy.

Improved genetics is just one of the tools for increasing farm production, but it is only effective if used in combination with other better farm management practices. This is described by animal geneticists as follows:

$$\text{Phenotype} = \text{genotype} + \text{environment}$$

The performance of milking cows (phenotype) then results from a combination of:

- genetic makeup (genotype), derived from the genes supplied by both dam and sire
- management in the milking herd (environment) – how well they are housed, fed and protected from climatic stress, disease and other stresses of dairy farming.

Domestic beef and dairy cattle are essentially of two types: tropical Zebu *(Bos indicus)* and temperate (*Bos taurus*). The temperate breeds have been intensively selected

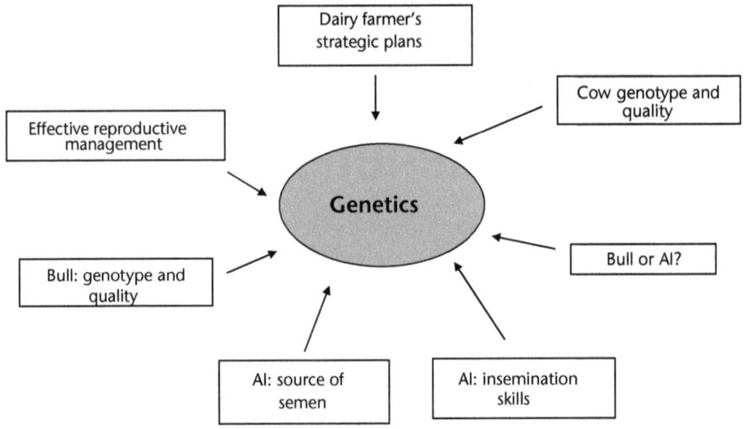

Figure 9.1. The key factors affecting genetic development of the milking herd

for meat and milk production for several centuries, whereas such breeding programs have only very recently been applied to the tropical breeds.

The relative importance of the genotype and environment differs between climatic zones, farming systems and farmer skills, and it is not easy to apportion relative values to either one. This is a major dilemma for dairy industry policymakers in deciding whether the long-term outcome of schemes to import high genetic merit dairy stock has been genetic improvement, herd multiplication or a combination of the two. Unfortunately, in too many cases, the genetic benefits of such importation schemes have been minimal. Much of the following information is over 20 years old, but many tropical dairy specialists would consider that little has changed and the same types of mistakes are being continually repeated.

9.1 Problems with exotic genotypes

The decision makers of most, if not all, Asian countries place high credence on improving the genetic quality of their national herds. This has led to an influx of *Bos taurus* dairy stock from temperate areas, where they have been selected for many generations. According to McDowell (1994), over 25 countries in the 'low latitudes' (namely equatorial and tropical areas) are developing local milk industries based on imported dairy cattle, mainly Friesians from the temperate developed dairy industries in Europe, North America and Australasia. During the 20 years to 1995, five million Friesians have been exchanged in international trade, of which half have moved from temperate climate countries to those in the warm climate zones. All countries in Latin America have nucleus herds based on Friesians, as do 70% of Asian and 66% of African countries. The major exception has been India, which has based its genetic improvement program on indigenous dairy cattle and buffalo.

The importing countries are also being encouraged, often through financial subsidies, to use semen from progeny tested sires from countries of origin of these cattle, thus continuing the program of genetic upgrading of the temperate dairy stock. But with

A proud farmer with his imported dairy heifer (Indonesia).

few exceptions, ages at first calving of 30+ months, milk yields of 2500–5000 kg/lactation and calving intervals of 430–485 days in the new location are below optimum and will not even support bank rates of credit (McDowell 1994). Modern day Friesians are genetically capable of producing their first calf at 2 yr of age, then one every 12 months, yielding 8000 kg milk over 300 days and milking for five lactations.

Ibrahim *et al.* (1992) reviewed 10 yr of dairy development in Indonesia (East Java), concluding that increases had been mainly through imported cattle, but productivity gains were only at one-third of their genetic potential. The main reasons they listed would be relevant to many SE Asian countries, being:

- lack of sufficient high-quality feed
- poor adaptability of imported stock to local conditions
- poor housing and management
- too much attention of cooperatives to marketing and insufficient extension on cow problems
- failure to repay dairy credit in view of low productivity and poor milk prices.

9.1.1 Jerseys versus Friesians in the tropics

Apart from small numbers of Brown Swiss and Red Danes, Jerseys and Friesians have been the exotic breeds of choice for genetic upgrading in the tropics. On the whole, Jerseys and their crossbreds have lower full lactation yields of milk or milk solids than Friesians and their crossbreds (Tibbo *et al.* 1994). The Jerseys' earlier age at first calving

A quality Friesian heifer selectively bred in Pakistan.

reflects their faster rate of maturity. Friesians have longer calving intervals, presumably because of their longer period of negative energy balance post-calving. For other production measures, such as services per conception, conception rates, abortion rate and level of mastitis infection, breed differences are small. Increasing numbers of dairy development programs in SE Asia are looking towards purebred and crossbred Jerseys to form the basis of their exotic breed imports. Being smaller and in less demand in developed countries, Jerseys can generally be sourced more cheaply and have lower transportation (air) costs than Friesians.

9.1.2 Genotype by environment interactions

Like all genetic improvement programs, be they livestock, cereal grains or fruit trees, the use of exotic gene stock needs to be accompanied with modifications to the environment. McDowell (1994) contends that this is not happening with dairy cattle because there is still the perception that Friesians, Jerseys or other improved dairy breeds can perform well on diets based on low to moderate digestible tropical grasses in a foreign environment with temperature, disease, pollution and other production constraints.

Small holder dairying uses a wide range of dairy stock in the humid tropics. These include the following types:

1. local, unimproved cattle with little breeding for milk production
2. crosses between local cattle and Zebu dairy breeds

3. crosses between local cattle and temperate dairy breeds
4. purebred temperate dairy breeds.

There is continuing discussion over the 'ideal' genotype for small holder dairying in the humid tropics. One rational approach is to consider the genotype by environment interaction (that is G × E). Briefly, the relative performance of two genotypes depends on environment under which they are managed. When upgrading a national dairy industry from type 1 above, is it better to base the industry on types 2 or 3 and eventually on type 4? More importantly, at what level of feeding and general herd management should the industry move from type 3 to type 4 cows?

The evolution of the dairy industry in the wet tropics of Central America is a good example of understanding how the breed must fit the system. There, the industry is now based on dual purpose cattle rather than, as it was 20 yr ago, on traditional specialist dairy breeds such as Friesians. Cows are milked with the calf at foot and weaning generally coincides with the end of lactation. Income is generated from milk plus quality meat from yearling stock, rather than just the poor-quality meat from cull cows as is usual in most dairy systems. This system has evolved because the dairy cows, being Zebu type, require the presence of a calf to stimulate milk let-down, and they tend to dry off prematurely when daily calf contact ceases, and because milk substitutes and high-quality calf concentrates are either not available or prohibitively expensive.

No single tropical dairy type can be defined as the 'best'. The 'best' type will vary from local genotypes through to high grade Friesians. Unless heat stress is controlled, very high-yielding temperate breeds will always be less efficient than locally evolved ones. Such animals would be developed through crossbreeding local and exotic breeds then selecting within these populations for high milk yields. Similarly, it is unlikely that purebred local breeds will be the most profitable in all but the poorest of situations, where feed is inadequate, parasites and diseases are uncontrolled and levels of production barely fulfil household needs.

The genotype and its management must then be matched to the:

- climatic conditions that exist
- available nutrition, whether home grown or supplemented with purchased feeds
- degree of challenge from parasites and diseases
- level of management skills
- availability and costs of labour to feed, milk and market the product, and its priority with other household demands
- availability of finance
- availability and access to profitable markets.

9.2 Specially bred tropical dairy genotypes

The ideal tropical dairy genotype is a small animal that yields high levels of milk and/or milk solids, produces a live calf (preferably a heifer calf) annually under simple small holder production systems based on tropical forages and with minimal environmental manipulation and low exposure to disease. Unfortunately, such a genotype does not exist because high milk yields require high intakes of quality forages together with high internal heat production, hence management to alleviate heat stress.

Over the last 50 yr, there have been various programs to develop tropically adapted dairy genotypes. These include the Jamaica Hope (80% Jersey, 15% Sahiwal and 5% Friesian) in the Caribbean and the Australian Milking Zebu (AMZ) and Australian Friesian Sahiwal (AFS) in Australia. The objective of such breeding programs was to breed a stabilised genotype with the combined attributes of tropical adaptation and superior milk and reproductive performance. Although such stock would not exhibit the hybrid vigour of crossbreeds, their progeny would be less variable in their physical and productive traits.

The AMZ program, which ceased in the 1970s, was based on Jersey and Red Sindhi breeds, and produced a relatively small cow with low yields of high solids milk. Such a genotype would be ideal for many small holder industries, such as in Indonesia, where most of the milk is destined for industrial processing (hence the benefits of high solids content) and the feeding management is better suited to small cows with low maintenance requirements. However, when recently asked about the suitability of such a genotype, industry leaders in Indonesia considered it less suitable to Jersey or AMZ because of the Friesian's better dairy beef sale value for bull calves or cull cows.

The more successful AFS program, which ceased in the 1990s, was based on Friesian and Sahiwal breeds. Such animals are currently in great demand throughout Asia because of their performance in small holder dairy systems. Their Sahiwal ancestry provided the desired tropical adaptation to complement the dairy performance of the Friesian. Preliminary discussions are now underway to resurrect this breeding program. Such a program would require a good database on the genetic variability of local Friesian populations: a shortfall in many Asian countries where robust herd recording, and performance and progeny testing programs are often lacking.

Another dairy breeding program creating interest in the tropics has produced the Brazilian Girolanda breed. These cattle are now the major dairy breed in Brazil and were bred by crossing the local Gir (*Bos indicus*) cattle with the Friesian. Although popular since the 1940s, the breed was officially established (containing 5/8 Friesian and 3/8 Gir) only in the late 1980s. Some of its attributes include:

- It is reputed to exhibit a degree of hybrid vigour as a result of its two vastly different parent breeds.
- It has good tropical adaptation, thus has good milk production in all tropical climates.
- It has a quiet temperament.
- It exhibits ease of calving.
- It has high fertility, particularly good conception rates during the hotter months.
- It has longevity in the milking herd.
- It has good resistance to parasites.

Another *Bos indicus* breed that has been selected for dairy production is the Sahiwal. This is an indigenous dairy bred from Pakistan and, when managed well, can be a good milk producer. The Sindhi is another potentially good Zebu dairy breed from the Indian subcontinent.

9.3 Breeding decisions to be made on the farm

Each dairy farmer should have a strategic plan for his breeding management. For SHD farmers, plans are likely to be very similar, with the major decision being the particular breed of milking cow the farmer desires and that the cow only requires a minimal number of services or artificial inseminations. Some farmers may like a particular bull type, whose progeny are more likely to have some specific body conformation or certain milk compositional characteristics, but small holder farmers rarely get to have much choice in the semen they can source.

Accessibility to AI is a major determinant of planning a breeding program. SHD farmers are generally limited in their choice of semen, but Friesians are usually the breed of choice provided by the inseminators working for the dairy cooperatives, although, when available, AFS semen is very popular. It may be worthwhile trying to source semen from other dairy genotypes, such as Jersey or Brown Swiss, to compare with Friesians. Advice from your semen supplier should be sought.

If AI is not available, the only choice is natural mating, and in many cases that requires each farmer to have his own bull. This limits the genetic improvement possible compared with AI because the bull is likely to be locally bred and of poorer genetic merit than bulls used to supply semen. Bulls are quite expensive for a small holder farmer to own, so, provided there are sufficient dairy farmers in the area, and the owner of a good quality bull is agreeable, using a common bull may become an option. If this is possible, it is important for a veterinarian to check the reproductive health of any herd bull.

Bulls are generally better than humans at picking out any cow on heat. Provided the bull is fertile, its reproductive performance is likely to be better than an inseminator, particularly if he only has a limited number of cows to inseminate in his area. Lack of potential inseminations each year can severely limit inseminator skills. Tropical climates can also reduce the ease of detecting when cows are on heat.

The impact of AI on the genetic improvement of national herds can be very low. For example, in Sri Lanka, despite the extensive use of AI, over 65% of the progeny result from natural services, mainly from scrub bulls. This is attributed to the poor infrastructure of AI services in the country, as well as the lack of farmer confidence that AI will result in a pregnancy.

There are several relatively new reproductive developments that SHD farmers may be aware of. Oestrus synchronisation is hardly relevant for the small holder, while embryo transfer is much too expensive to consider. Single sex semen is still expensive but in years to come, it may become sufficiently cheap to be an economic option.

Genetics is all too often given too much emphasis in the future development of SHD herd-management programs. Nutrition is by far the greatest constraint to farm milk production and profits. For AI to have much impact on the genetic improvement in many developing countries, farmers and government or dairy cooperative inseminators require considerably more training in reproductive management.

10

Managing stock surplus to the milking herd

This chapter discusses the classes of stock on the dairy farm that are sold to generate income.

The main points in this chapter

- Milking cows can be culled because of poor performance, chronic animal health problems, such as mastitis, or acute injuries, such as broken limbs.
- Bull calves are frequently sold as week-old animals.
- Dairy beef from bulls or steers, particularly purebred or crossbred Friesians, can be quite profitable.
- Their feeding management is similar to that of dairy heifer replacements, which was discussed in Chapter 4.

Small holder dairy farms can generate income additional to the sale of raw milk and value-added dairy products. These include:

- cull cows
- excess dairy stock, such as bull calves
- dairy stock specifically grown out for beef, called dairy beef
- manure
- excess fodder
- grass cuttings for planting material on other dairy farms
- biogas, from dairy effluent.

This chapter deals specifically with managing stock surplus to the milking herd.

10.1 Cull cows

Milking cows are normally sold when the farmer considers them uneconomic to try and mate to produce milk in future years. For example, they could be culled from the milking

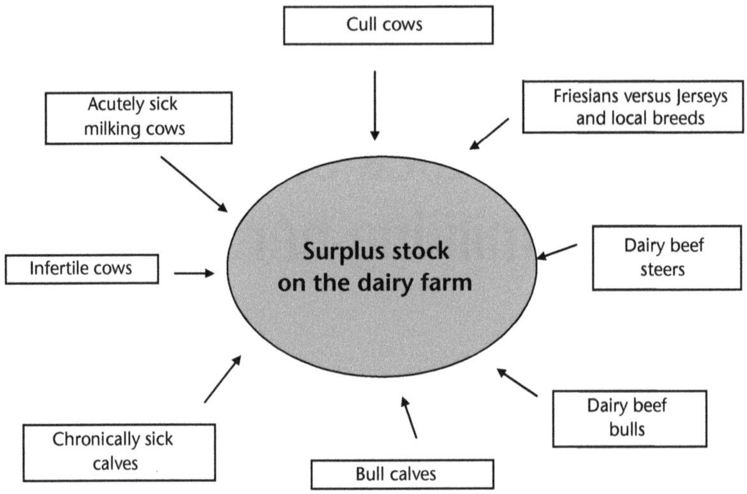

Figure 10.1. Surplus stock on the dairy farm

herd because of acute animal health problems, such as acute trauma or acute sickness. They could also be culled because of chronic disease problems, such as infertility or mastitis. They may be selected for sale because their milk production is minimal. They may be sold because the farmer needs the space or the feed for more productive stock. He may also need additional finances to purchase other stock or for personal reasons, such as a family celebration.

If the cull cows show symptoms of acute disease, care must be taken to reduce the likelihood of that disease spreading to other stock on the farm. They should be isolated from all stock and when transported to the saleyards, the vehicle should be thoroughly cleaned and sterilised. The major objective should be to remove them from the farm as rapidly as possible.

It is unlikely that farmers will want to improve the body condition of any cows culled because of chronic animal health problems, although the farmer may want to keep them on the farm if their sale price is likely to improve in the foreseeable future. If they are still milking, their diet should be the same as other milking cows and, for say infertile cows, they can be milked for as long as the farmer considers their returns from milk are greater than their feed and other farm costs they may incur. If non-lactating, they should be fed sufficient nutrients to at least maintain their body condition, so their sale price is still maintained.

If cows are culled because of acute injuries, such as a broken limb or dislocated joint, they should be sold quickly and with as little additional stress as possible to ensure the animals' welfare is not compromised. Farmers should, and mostly do, fulfil their obligations to the wellbeing of their livestock.

Another class of cull stock are young calves that have a history of chronic or acute health problems and are unlikely to ever become productive cows in the herd. Such decisions are value judgements, but there is much evidence that such heifer calves are likely to be culled from the milking herd much earlier in their life than would calves that did not suffer the growth setbacks in their early life. It is worthwhile recording which

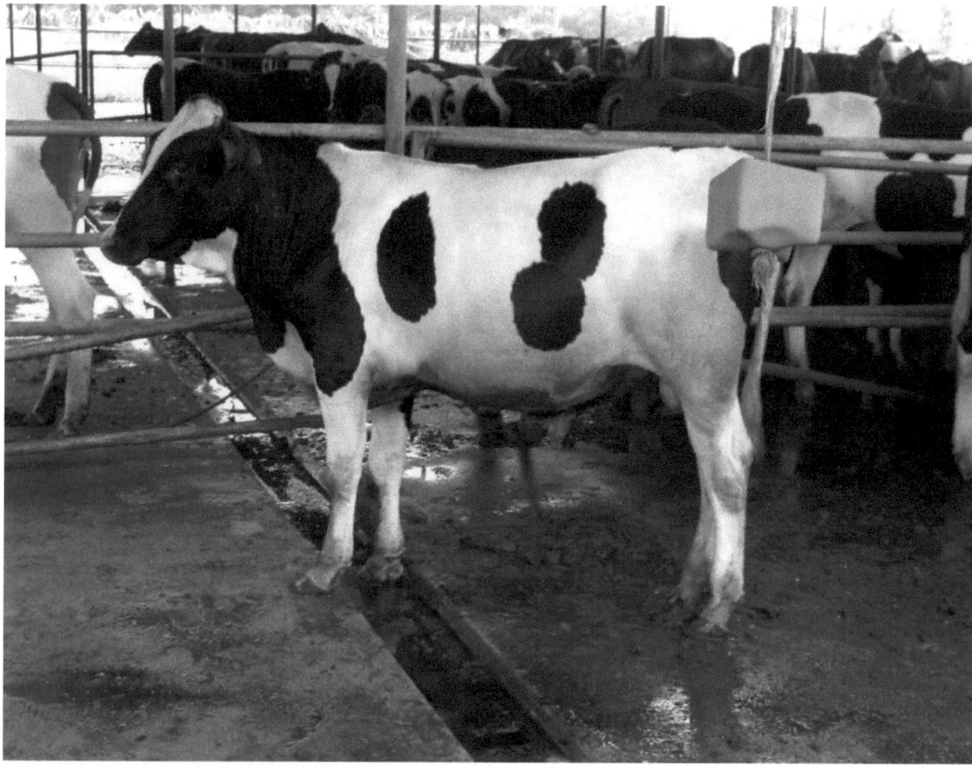

A Friesian herd bull bred for sale in Malaysia.

calves suffer from chronic disease in early life and noting their longevity in the herd. This information will assist farmers with their value judgements as to which of these calves should be culled. Selling them just passes the problem onto other farmers who may want to buy them, so it would be better to euthanase them rather than transport them to the saleyards.

10.2 Excess dairy stock

Low-producing cows would also be classified as excess dairy stock, but this class refers more specifically to male calves and potentially poor performing heifer calves. These are normally sold within a week or so of birth.

In Australia, animal welfare regulations stipulate that these animals must be more than 4 days old and should weigh more than 23 kg. Newborn calf age is usually determined by the state of the umbilical cord, or the string as farmers frequently call it. If it is moist, it is likely that the calf is less than 4 days old.

10.3 Dairy beef

Growing out bull calves for slaughter at 18 months or more, is frequently a major generator of income for dairy farmer. These stock can be kept entire and grown out as

Growing out Friesian bulls for sale (Indonesia).

bulls, or they can be castrated at an early age and grown out as steers. Bulls grow faster than steers but, when they reach puberty, they can become a nuisance with milking cows on heat. Feeding management for dairy heifer replacements was discussed in Chapter 4, and these guidelines could also be used for growing out dairy beef bulls or steers. Target growth rates should be at least 0.5 kg/day. There may be additional recommendations with regards to vaccinations and other health management issues, and such information should be sought from local animal health specialists.

Purebred or crossbred Friesians are preferred over Jerseys or local breeds and generally return more money per kg live weight when sold for dairy beef, so the breed of the bull calf should be considered when deciding on its future fate on the farm. On the whole, milk generates more income per kg fresh grass grown on the farm than does beef, so it is important to 'do the sums' on dairy beef production when farm stocking rates are high. Acceptable growth rates are unlikely if the major part of the ration is very low-quality forages, such as rice straw. As with milking cows, green grass is the best forage to feed dairy beef animals.

It is important to plan the finishing (or fattening) strategy for dairy beef. Target live weights should be used in any feeding program, so using a chest girth tape to estimate live weight is worth considering. Friesians are later maturing than other dairy beef animals, meaning that a higher level of concentrate feeding will be required to produce a suitable sale animal: namely one with a good degree of finish (or some cutaneous fat cover over the ribs).

10.4 Imported cows in lactation anoestrus

Another class of surplus stock are healthy dairy cows that do not cycle. This can be due to poor feeding management during early lactation and, unfortunately, this is becoming more common with imported high grade dairy heifers. Such animals may be imported as pregnant heifers, have their calf and then fail to cycle. Their genetic propensity to preferentially use body reserves to produce milk means that they are in continual negative energy balance on farms with very poor feeding management, hence suffer lactation anoestrus. In some tropical regions, these can represent up to 20% or 30% of imported dairy stock, which have been imported at very high cost. This makes for very expensive dairy beef!

11

Stock welfare

This chapter discusses the key elements of animal welfare for SHD farmers.

The main points in this chapter

- There are five basic freedoms for livestock: freedom from hunger and thirst, from discomfort, from pain, to express normal behaviour, and from fear and distress.
- It is important that facilities and equipment are well designed and maintained.
- Stock should always be provided with adequate feed and water.
- Herd management practices should always minimise stress, injury and disease.
- If necessary, stock should always be humanely euthanased.
- Staff should be adequately trained or experienced in good animal care.
- Special attention should be given to stock selected for sale and transport.
- Calves are very susceptible to stress, injury and disease, so management practices should be undertaken with this in mind.

Throughout the world, public perceptions of farm animal welfare issues have the potential to markedly affect the sustainability of livestock industries, with national and international pressures likely to have increasing roles in determining how animals are managed. Because farm animal welfare is largely part of good animal and farm management, paying close attention to their day-to-day management should also ensure acceptable welfare.

11.1 The five basic freedoms of livestock

The welfare of cattle can be summarised using the 'five basic freedoms' as follows:

1. Freedom from hunger and thirst, through ready access to fresh water and a diet to maintain full health and vigour.
2. Freedom from discomfort, through provision of appropriate shelter and comfortable resting areas.
3. Freedom from pain, by prevention and, when sick, rapid diagnosis and treatment.

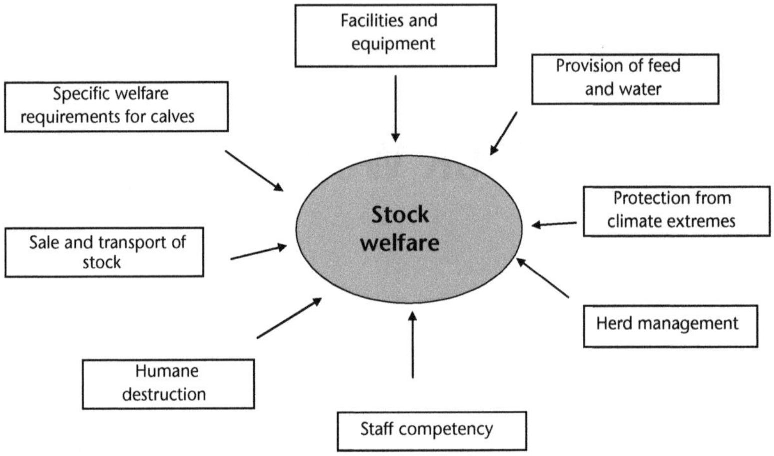

Figure 11.1. Considerations for stock welfare on the dairy farm

4. Freedom to express normal behaviour by providing adequate space, proper facilities and the company of other animals.
5. Freedom from fear and distress by ensuring conditions and treatment that avoids suffering.

All management and housing systems should be designed, constructed, maintained and managed to assist with these five freedoms.

11.2 The six basic elements of stock welfare

Specific to dairy herd management, Edge *et al.* (2006) have broken dairy welfare down to six basic elements as described in sections 11.2.1 to 11.2.6.

11.2.1 Facilities and equipment

These should be well designed, maintained and operated to ensure a high level of animal welfare, minimal stress or chance of injury. This includes:

- protection from extremes of weather
- provision of sufficient space with minimal possibility of injury
- flooring designed to minimise slipping, falling and lameness
- milking machinery maintained regularly to ensure minimum stress to cows
- facilities for water, feeding, milking and restraint that are appropriately designed and maintained
- laneways, tracks and gateways designed to minimise stress and injuries
- equipment for euthanasia clean and in good working order.

11.2.2 Provision of feed and water

These should be provided with consideration of environment, age, body condition and stage of lactation that ensures their health, wellbeing and productivity. That is:

- Stock should have access to feed and water of appropriate quality and amount.
- Stock should be fed rations formulated and balanced to provide the necessary nutrients for their desired level of performance.
- Short- and long-term plans are in place to manage feed and water shortages and drought.
- Practices should be in place to reduce the potential risk of toxicity or contamination during storage.

11.2.3 Herd management

Dairy cattle should be routinely managed and handled to minimise stress, injuries and disease and promote good health and welfare. Such procedures should ensure that:

- Reproductive practices should be carried out competently to ensure good animal welfare and reproductive outcomes.
- Other routine husbandry procedures, such as calving induction, dehorning and castration, if required, are carried out to minimise pain.
- Systems to manage disease and other animal health disorders should optimise the planning, prevention and monitoring of dairy herd health.
- Disease, injury, illness and stress should be identified and treated promptly, with suitable expert advice sought as required.
- Animals are regularly inspected to monitor their health and welfare.

11.2.4 Humane destruction

Weak, ill or injured cattle and calves should be identified and treated appropriately or humanely destroyed using approved methods:

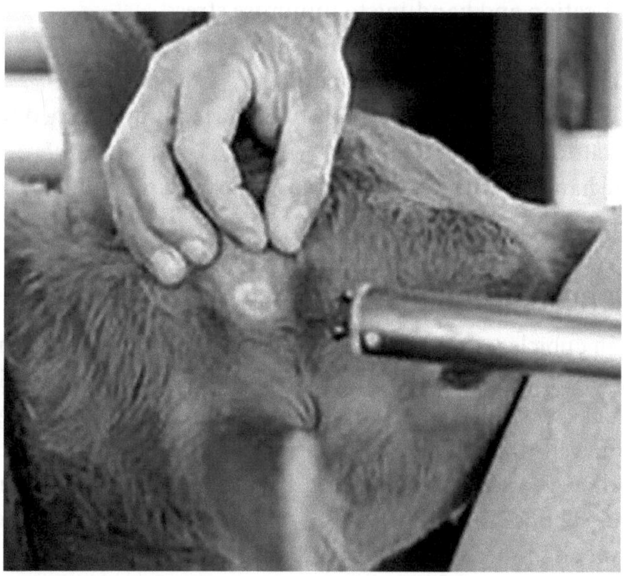

Dehorning of calves should take place within their first 2 months.

- Stock requiring humane destruction should be identified and promptly euthanased.
- Competent staff must be available to carry out humane destruction, using approved methods.

11.2.5 Staff competency

All staff responsible for managing and handling dairy stock should be competent in their tasks and aware of their responsibility for good animal care. That is:

- Staff carrying out routine husbandry, surgical procedures, reproductive procedures, administering health treatments and handling and transport of cattle must be appropriately trained or experienced.
- All staff handling stock must be able to identify signs of illness, abnormal behaviour or stress and ensure appropriate action is taken.
- A competent person must either be available on site or can be contacted to handle emergencies and humane destruction when necessary.
- Staff must behave in a manner to minimise fearfulness in cattle.

Stockmanship, plus the training and supervision necessary to achieve required standards, are the key factors in the handling and care of livestock. A management system may be acceptable in principle, but without competent, diligent stockmanship, the welfare of animals cannot be adequately safeguarded.

11.2.6 Preparation, selection, sale and transport of stock

Stock should be selected and appropriately prepared for transport to ensure they are fit for the intended journey:

- Preparation for sale and transport includes appropriate actions for feed and water curfews, identification and handling to minimise stress.
- Stock must be selected for transport with consideration of age, class and condition to ensure they are fit for the intended journey.
- Any weak, ill or injured stock must not be transported until deemed fit by a competent operator.
- Stock should be handled with care during loading and unloading.

11.3 Other general principles of stock welfare

Whether housed in cubicles, straw yards or cow sheds, in order to maximise performance and ensure satisfactory standards of welfare, the accommodation must provide for the stock's basic needs. As an absolute minimum, the housing must provide a comfortable, clean, well-drained and dry lying area, together with shelter from adverse weather. It must allow the animal to move freely around without risk of injury.

Cows at pasture choose to lie down for 12–14 hr each day, so a similar target should be achieved with stock in sheds. If cows spend less time lying down, they are likely to spend more time standing in loafing or feeding areas, which can adversely affect hoof

Tethering of stock in the sun can become an animal welfare issue (Pakistan).

health. Sheds should provide a minimum loafing area of 3 m^2 per cow with at least 3 m between rows of cubicles.

Most farmers believe that animal welfare is just good cattle husbandry. The major welfare issues facing dairy farming include:

- housing and cow comfort
- castration
- dehorning
- branding
- transportation
- slaughter.

Lameness can be a major problem, both from the point of animal welfare and farmer profits. Farmers frequently only recognise 40–50% of lameness problems and these are often well advanced, making them difficult to treat. The cow's environment, both social and physical, and her ability to cope with it ultimately determines how bad lameness can become.

Early separation of calves from their dam is another welfare issue that is practised for animal health reasons, to improve the likelihood of calves receiving sufficient quality colostrum in their first 24 hr of life.

It is important to maintain animals in good body condition, house them in social groups in clean environments with adequate space to move around and rest comfortably, maintain feed, medication and production records and compassionately handle animals undergoing management practices, such as dehorning or castrating, and those that are injured or ill.

Most Western countries have strict codes of animal welfare covering transport, housing, handling and feeding. Farmers purchasing dairy stock from such countries may be expected to abide by the principles of that country's animal welfare codes. This is particularly the case with young stock, with the following section detailing Australia's welfare code for dairy calves.

11.3.1 Problems of confinement

Dairy stock imported from Western countries have almost invariably been reared under grazing conditions, hence have never been exposed to a continual shed environment, as is common on most small holder farms. Compared with grazing, confinement creates specific problems such as:

- restricting opportunity to seek comfort; for example, if only provided with cement floors
- creating problems of high humidity, which can be more detrimental than high temperature
- limiting opportunity for exercise, hence the need for routine hoof trimming
- increasing exposure to infectious diseases
- other health issues, such as mastitis and uterine infections, when hygiene is poor during milking and calving
- creating problems of heat detection for artificial insemination
- requiring greater sanitation
- magnifying problems of social dominance
- increasing capital investment

Many of these issues are covered in other chapters in this manual.

11.4 Australian welfare codes for calves

Of all the classes of dairy stock, the welfare of milk-fed calves has received the closest scrutiny. The following summarises the current codes for their health welfare and wellbeing (Moran 2002).

11.4.1 Housing

- Housing should be hygienic, with adequate ventilation, climate control and lighting. Flooring should be well drained with adequate dry lying space for each cow. Flooring and internal surfaces should not cause injury and should allow easy cleaning.
- Calves should be given access to sufficient bedding or appropriate flooring that ensures comfort and a clean, dry place to lie.
- Feeding systems should be designed to permit easy access and reduce bullying.
- The floor area must be sufficient to enable each calf to freely turn around, stretch and lie down comfortably. A floor area of at least 1.5 m^2 should be provided for each calf individually housed in pens or cribs. Pen heights should be a minimum of 1 m with provision of additional height to allow for ventilation space.

- Calves, being social animals, seek the company of other calves, so should be able to see other calves during early rearing.
- Where large numbers of calves are reared, they should be grouped by age and size to reduce competition for food and to allow closer observation and management.

11.4.2 Feeding

- Calves require at least 2 L fresh or preserved colostrum within the first 12 hr following birth. Calves should continue to receive colostrum for the 3 days after birth.
- Thereafter they should be fed at least daily on liquid milk, or milk replacer, in sufficient quantities to provide essential requirements for maintenance and growth.
- Hay, calf concentrate or high-quality forage should be available to calves no later than 3 weeks of age to help in the development of their digestive tract and to ease the stress of weaning.
- Calves should be weaned off milk or milk replacer onto rations providing all essential requirements only when their ruminant digestive systems have developed sufficiently to enable them to maintain growth and wellbeing.

11.4.3 Management practices

- Restraint should be the minimum necessary to perform management procedures efficiently.
- Procedures and practices that cause pain should not be carried out if painless and practical methods can be adopted to achieve the same result.
- Special care should be given to the practices of castration, tail docking (if undertaken) and dehorning.
- Any injury, illness and distress should be promptly treated.
- Appropriate preventative measures should be implemented for diseases likely to occur in the herd. A suitable vaccination, and internal and external parasite control plan should be devised and followed for each farm.
- Internal medication, such as vaccines and drenches, and external medications, such as dips and pour-on formulations, should be stored and given in strict accordance with the manufacturer's instructions and recommended methods of administration. Overdosing may harm cattle while underdosing may result in failure of the medication. Expiry dates and withholding periods should be strictly observed.
- The preferred method of humane destruction (euthanasia) of cattle should be by overdose of anaesthetic or using a gunshot.
- Killing day-old calves may also be achieved by a heavy blow to the crown of the head to stun the calf prior to bleeding out.

11.4.4 Transportation of calves

- Transportation of stock should ensure that they reach their destination as speedily as possible, within the confines of the road law, and in a condition not significantly less than when assembled for loading. The possibility of either injury or illness during transport should be reduced to a minimum.

- Handling of calves should be carried out in a manner that will avoid injury or unnecessary suffering. Calves are not to be kicked, beaten, pulled, thrown or prodded with any sharp instrument. The use of electric goading devises or dogs when handling, driving, drafting, weighing, loading or unloading is not an acceptable practice.
- Facilities should be constructed to permit safe loading and unloading of calves.
- Places where calves are held should have facilities and/or contingency plans to feed calves in the event of delayed removal or slaughter.
- Calves should be fed at least every 24 hr and have access to drinking water.
- The driver of the vehicle is responsible for the care and welfare of all animals during transportation.
- Animals that either become ill or weak, or are injured during transport should receive appropriate attention and treatment; if necessary, they should be slaughtered humanely.
- Vehicles used for transportation should be thoroughly cleaned prior to loading and at the end of the journey.
- Transport operators should check calves en route at least every 3 hr.
- Calves should be loaded at a density so as to allow all calves to lie down while being transported.

12

Environmental management

This chapter discusses the principles and practices to minimise the adverse effects of heat stress on the performance of dairy stock.

The main points in this chapter

- The comfort zone of milking Friesians is 6–18°C. Outside this range, stock must modify their behaviour, their thermoregulatory processes or reduce milk production and fertility.
- The severity of heat stress depends on many factors such as diurnal temperature, housing conditions, breed and level of milk production.
- The Temperature Humidity Index is a quantitative comfort index that relates closely to cow performance.
- Respiration rate is a simple measure of heat stress that farmers can easily use to modify management if necessary. This has been developed further into a panting score.
- Cooling measures for milking cows include designing sheds for maximum ventilation, sprinklers, fans, allowing cows outside during the evening and modifying feeding management.

12.1 Heat stress in milking cows

The comfort zone for milking Friesian cows is 6–18°C. Within this range, there are no measurable fluctuations in their physiological processes, while the energy input to output shows good biological efficiency, in that all body processes will be functioning in their expected ranges. Between –5°C and +5°C, appetite will be stimulated, while at the upper level, above 27°C, appetite is depressed and both biological and economic efficiencies decline. Above 24°C, feed intakes decreases by about 3% for every rise of 1.2°C.

The extent of the effects of temperature on appetite depends on the:

- type and quality of forage; intakes of high-fibre forages are more depressed at high temperatures
- type and quality of concentrates; cows will selectively eat more concentrates and less forages when heat stressed

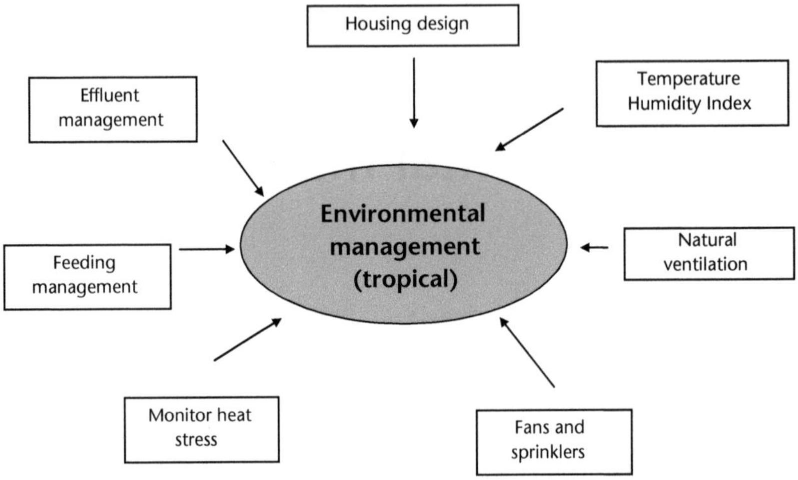

Figure 12.1. The key elements of environmental management on tropical dairy farms

- humidity; high humidity exaggerates the effect of high temperature
- stage of lactation; cows in early lactation are more susceptible to heat stress
- milk yield; high-yielding cows are more susceptible to heat stress
- actual appetite.

The basic thermoregulatory strategy of cows is to maintain a body core temperature higher than ambient temperature to allow heat to flow from the animal via four basic routes of heat exchange, namely:

- conduction, in which heat moves from a warmer to a cooler surface, so cows need direct contact with the surface
- convection, in which the layer of air next to the skin is replaced by cooler air
- radiation, in which heat can radiate from a warmer to a cooler environment
- evaporation, in which sweat or moisture is evaporated from the skin or respiratory tract.

The first three routes require a thermal gradient to operate in which air temperature must be lower than body temperature. Once air temperatures approach body temperature, the only viable method of heat loss is evaporation. This requires a vapour pressure gradient to be effective, so is very dependent on the humidity, or amount of moisture in the air.

Because cattle have a limited ability to sweat, the main route of heat loss in cattle during hot weather is evaporative cooling from the respiratory tract, namely the nasal passages and lungs. Cattle then increase their breathing rate to increase movement of air over the moistened surface of the upper respiratory tract and mouth. However, if humidity levels are high, the effectiveness of this evaporative cooling is decreased and cattle may be unable to dissipate accumulated body heat.

When planning feeding programs, consideration should be given to the number of hours each day when temperatures exceed 27°C and relative humidity exceeds 80%; feed intakes will decline once temperatures exceed 27°C or higher for 6 hr. Furthermore, high body temperatures reduce the efficiency of rumen digestion and increase body maintenance requirements, further increasing the energy deficit. The net effect on feed intake depends on the number of hours each day below 20°C, which allows cows to cool, hence restore their heat balance.

12.1.1 Symptoms of heat stress

There are many symptoms of heat stress (those more relevant to shedded cows are shown in italics) (Moran 2005). The initial signs are behavioural, while the last five signs are the more severe physiological ones due to a failure to cope, hence require immediate attention to reduce their adverse effects on cow performance. In order of increasing severity, they are:

- body aligned with direction of solar radiation
- seeking shade
- *refusal to lie down*
- *reduced feed intake and/or eating smaller amounts more often*
- crowding over water trough
- body splashing
- *agitation and restlessness*
- *reduced or halted rumination*
- grouping to seek shade from other animals
- *open mouthed and laboured breathing*
- *excessive salivation*
- *inability to move*
- *collapse, convulsion, coma*
- *physiological failure and death.*

As well as behavioural symptoms, heat-stressed cows will produce milk containing less milk protein or solids-not-fat. In addition, milk fat levels may decrease if cows markedly reduce their forage intakes.

The severity of heat stress depends on many factors. These include the :

- actual temperature and humidity
- length of the heat-stress period
- degree of night cooling that occurs
- ventilation and air flow
- cow breed and size
- level of milk production and dry matter intake prior to heat stress
- housing type, overcrowding and aspect
- availability of water
- coat colour, if exposed to sun
- hair coat depth.

Milking cows are maintained in a variety of environmental conditions. Without access to shade, the heat load on cattle grazing at pasture is generally lower than for cattle in dirt yards, because the dirt surface absorbs less heat then grass, thus radiates more heat onto the stock. For example, the surface of a dirt yard can reach 60–80°C (on a day with high solar radiation and ambient temperatures of 40–45°C), but it will cool down rapidly once the sun sets. Clearly, access to shade, whether at pasture or in yards, is highly desirable in regions with high radiation heat loads.

In addition, heat stress adversely affects reproductive performance in three ways:

- Acute stress can lead to embryo reabsorption, while chronic stress upsets normal cyclic status, through hormonal changes, particularly if cows are exposed for 6 hr or more to temperatures above 27°C.
- In late pregnancy, reduced foetal growth can also result from heat stress, leading to increased calf mortalities.
- The intensity of expression of oestrus is depressed, in that oestrus periods are shorter (for example 12 versus 17 hr) and, although cows do cycle during hot periods, the percentage of those actually observed can be as low as 35–40%. This can be partly overcome by more frequent observations, such as every 6 hr rather than 12 hr.

Most of the heat stress management research has been conducted with Friesian cows producing 25 kg milk/day or more. Many of the Friesians in Asia produce only 10–15 kg milk/day at peak yield. Such low levels of milk production are due primarily to poor feeding and water management. Poor feeding management involves the feeding of low-quality roughages, which will exaggerate heat stress because such fibrous feeds produce a lot of internal body heat during their digestion. Poor water management occurs when milking cows are rarely offered free access to drinking water and what volumes of water they are offered generally restrict feed intakes as well as water available for cutaneous evaporation.

Therefore one should ask how relevant are the heat stress guidelines for high-yielding cows to Asia's situation? The answer is that we really do not know until the relevant research has been undertaken. The data are just not available to develop heat stress guidelines for other temperate dairy breeds, such as Jerseys, let alone tropical dairy breeds. In the meantime, using the published guidelines for environmental management would be the best bet. Because reproductive efficiency is more sensitive to heat stress than is milk production, and poor calving performance is an all too common feature throughout Asia, current shed designs and cow cooling practices are clearly inadequate.

12.2 The Temperature Humidity Index

The best single descriptor of heat stress is the Temperature Humidity Index (THI), because this combines temperature and relative humidity into a single comfort index. The higher the index, the greater the discomfort, and this occurs at lower temperatures for higher humidities. The THI is presented in Appendix 1 of this manual, while its effect on cow performance is summarised in Table 12.1.

Heat load index

MLA (2006) has developed a heat load index to assess environmental heat load on feedlot beef cattle. This is based on a combination of measures of heat load, namely:

- black globe thermometer – a measure of radiation heat load that takes into account both ambient temperature and solar radiation

Table 12.1. Effects of Temperature Humidity Index (THI) on dairy cow performance

Comfort zone	THI	Stress	Comments
A	<72	None	–
B	72–78	Mild	Dairy cows adjust by seeking shade, increasing respiration rate and dilution of blood vessels. Cow performance is adversely affected with reproduction more so than milk yield.
C	78–89	Severe	Both saliva production and respiration rates increase. Feed intakes decrease while water intakes increase. Milk production and reproduction are both reduced.
D	89–98	Very severe	Cows will become uncomfortable due to panting, high saliva drooling and high body temperatures. Milk production and reproduction will markedly decrease.
E	>98	Danger	Potential cow deaths can occur

Comfort zone: A, No stress; B, Mild stress; C, Severe stress; D, Very severe stress; E, dead cows.

- relative humidity
- wind speed.

The index includes a number of adjustment factors such as genotype, coat colour, access to shade, water temperature in drinking troughs and whether the animal is sick or healthy. Use of this index over time allows for the calculation of an accumulated heat load and the required heat loss during the night to maintain zero heat balance. Developing such a heat load index for Asian small holder dairy cows, normally maintained in sheds, is unlikely to provide an additional useful management tool.

12.2.1 Adverse effects of heat stress

For Friesians producing 20 kg milk/day, a THI above 78 leads to a decline in milk yield. A THI of 78 occurs at 29°C with 50% humidity or at 27°C with 80% humidity. There is also a decline in milk composition (milk fat and milk protein contents), but this occurs at 1–2°C higher than corresponding break points for milk yield.

With regards reproduction, this declines before milk yield: namely at THI of 72, equivalent to 25°C plus 50% humidity or 23°C plus 80% humidity. Cows in early pregnancy (up to 3 weeks) can abort, while cows in mid-pregnancy can have reduced birth weights. Cows are also more likely to have shortened and/or silent heats (less than 8 hr). Heat stress delays heat (hence submission rates) and, at the time of insemination or during the following 3–5 weeks, it can reduce conception rates and increase embryo mortality. By comparing conception rates between seasons (hot versus cool or wet versus dry), heat stress may be diagnosed as a problem if seasonal conception rates differ by more than 10–12%.

Cows are particularly vulnerable at temperatures above 30°C or, above 25°C with high humidity. Cows producing more than 15 kg/day of milk are more susceptible to heat stress due to their higher metabolic heat load. Zebu cows are less susceptible than Friesians because of their dense flat coat and higher density of sweat glands, but exactly how less susceptible has not been documented. When planning strategies to minimise heat stress, it is important to give priority to non-pregnant cows, usually in early lactation.

The adverse effects of heat stress are delayed by several days. The effect of mean THI 2 days earlier has the greatest influence on milk yield, while the effect of mean temperature 2 days earlier has the greatest influence on feed intake.

Another good 'rule of thumb' when assessing heat stress for dairy cattle is that air temperature (in °C) added to humidity (in %) should be below 90.

12.3 Management practices to minimise heat stress

12.3.1 Clinical signs of heat stress

The following signs can be used to assess the degree of heat stress:

- mild heat stress: drooling, increased respiration to 80–100 breaths/min
- moderate heat stress: drooling, respiration of 100–120 breaths/min and occasional open mouth panting
- severe heat stress: drooling, respiration rate greater than 120 breaths/min and open mouth panting with tongue out. Cattle also have an agitated appearance, hunched stance and will often have their head down.

Cattle can move from mild to severe heat stress very quickly, within 30 min to a few hours. Therefore extra vigilance is required once mild heat stress is detected.

12.3.2 Monitoring respiration rates

Observing the behaviour of cows is important in deciding when to modify management. If respiration rates reach 70 breaths/min, milk yield and reproduction may be compromised; this corresponds to 39°C body temperature, in contrast to a normal body temperature of 38.5°C. Higher yielding cows have faster respiration rates, because of the extra body heat production associated with higher feed intakes and milk yields. For such animals, if respiration rates exceed 80 breaths/min in 70% of the cows, it is indicative of heat stress. Certainly, when they exceed 100 breaths/min, cooling strategies should be introduced.

Respiration rates are easy for farmers to monitor. The farmer should ensure the cow is standing or lying in a relaxed state and preferably cannot see the farmer. To improve accuracy, the farmer could move his hands in time with abdominal movements until they are at a steady rate. Using a watch, the farmer should count the abdominal movements for 10 seconds, repeating the exercise to ensure the count is consistent. Multiplying this by six will give the respiration rate in breaths per minute.

Monitoring respiration rates at various times of the day is a useful tool in assessing the suitability of sheds for milking cows. If rate exceed, say, 60 breaths/min in the morning, prior to the shed heating up, it is likely that the cows would benefit from simple modifications in their environmental management. It is unlikely that major modifications in shed design could be justified, such as increasing roof height or pitch or shed height at the side, although serious consideration should be given to constructing roof vents. If minor improvements cannot be made in the shed's natural ventilation, such as removing obstructions to the prevailing breeze, fans and/or sprinklers should be installed.

Table 12.2. Panting score to quantify heat stress

Panting score	Breaths/ minute	Breathing condition
0	<40	Normal with no panting. Difficult to see chest movement.
1	40–70	Slight panting, mouth closed with no salivation. Easy to see chest movement.
2	70–120	Fast panting with salivation present. No open mouth panting.
2.5	70–120	As for 2 but with occasional open mouth panting. Tongue not extended.
3	120–160	Open mouth panting and some drooling. Neck extended and head usually up.
3.5	120–160	As for 3 but with tongue out slightly, occasionally fully extended for short periods and excessive drooling
4	>160	Open mouth with tongue fully extended for prolonged periods and excessive drooling
4.5	Variable	As for 4 but head down. Cattle breathe from flank. Drooling may cease.

One enterprising farmer in Vietnam constructed a small shelter away from the cow shed, which maximised natural ventilation through a high roof and its location, making best use of prevailing wind. Whenever he noted cows with high respiration rates, he hosed them down then moved them to the small shed to alleviate their heat stress.

Panting score

The Australian beef industry has developed a panting score for use with feedlot beef cattle (MLA 2006). This could also be used by dairy farmers to asses heat stress in milking cows. Table 12.2 summarise the key features of the panting score.

Guidelines for lot-fed beef cattle are as follows:

- If more than 10% of the cattle exhibit panting scores of 2 or more, cattle handling should cease and only resume when conditions become cooler and cattle have returned to normal.
- Cattle with panting scores of 3.5 or more are in danger of death.
- If more than 10% of the cattle exhibit panting scores of 3.5 or more, there is potential for a serious problem to develop unless measures are taken to cool the stock.
- The transition from 2.5 to 4.5 can happen quickly – in less than 2 hr – under extreme conditions.

The panting score was developed primarily for lot-fed beef cattle often kept outside in dirt yards, with or without access to shade. This is not the normal situation with milking cows, which are either grazing at pasture (where heat loads are not as extreme as in dirt yards) or are maintained in sheds. Furthermore, the type of heat load is often different. In Australia, feedlot beef cattle are usually maintained in dry regions with high solar heat loads, whereas Asian dairy cows are usually kept in sheds where the heat load is more from high humidity and often poor air movement. However, there are some dairy regions in Asia, such as Punjab in Pakistan, where dairy cows are subjected to the same type of environmental management and heat stress as Australian lot-fed beef cattle.

12.3.3 Shed design

Assuming the sides of the shed are open to allow maximum ventilation, greater use can often be made of the principles of air movement when designing the roof. Because hot air rises, and considerable heat is produced by a concentration of milking cows in a shed, there should be an opening along the top of the roof, with a cap over it to restrict rain entering the shed. The roof slope should be greater than for feed sheds, namely 3–4° per 2.3 m, with the opening at least 50 cm wide, and the full length of the shed. A second design is a roof slope of 33° (4 in 12), with a vent at the top of 30 cm plus 50 mm per 3 m of width for sheds more than 6 m wide. The lowest point of the roof should be 3 m from the ground. The steeper roof pitch increases air flow across and above the roof, thus creating negative pressure over the opening. This hastens the flow of air out the top, as well as creating turbulence of air movement around the cows. If farmers are concerned about rain entering the shed through the vent, a gutter system can be installed below the ridge opening, while the concrete floor can be sloped away from the feeding area.

The shed should be sited so that wind breezes are not blocked by any obstacles or other buildings. Ideally, it should be on the highest ground possible, which will also be good for drainage of effluent, with other buildings located downwind on the site. There should be a minimum of four times the height of the nearest wind barrier as a horizontal separation. The ideal orientation, from a ventilation point of view, would allow the prevailing winds to hit the shed perpendicular to the side. This allows the wind to travel the shortest distance before exiting the shed, to improve the rate of air exchange and provide the cows with fresh air. The longer the shed, the more important is this perpendicular orientation to the prevailing winds. Other factors to consider are exposure of the outside stalls to sunshine, future expansion plans, cow flow, traffic flow and manure flow.

A north–south orientation is preferable to allow the sun to dry underneath both sides of the shed. An east–west orientation is not recommended because the southern side will have less opportunity to dry out. Trees should be planted on the western side of the shed to reduce solar radiation. Shade cloth, which blocks 80% of the light, can also provide protection provided it does not interfere with ventilation within the shed. I have seen mosquito netting across many shed openings in some regions, but such a decision would depend on recommendations of local veterinarians as to the potential animal health dangers of mosquito infestations. Eaves that extend one-third the side height will provide good sun protection.

White painted buildings reflect solar radiation better than dark painted buildings. Reflecting roof materials such as galvanised or aluminium are good long-term investments. Insulation under the roof can reduce the heat load, and one cheap and effective method is to use old egg cartons from poultry farms because these create large air gaps in the roof space. Spraying water on the roof may only be effective in reducing roof and shed temperatures in areas with low humidity.

12.3.4 Cooling cows

Garden hose

The easiest method to cool a cow is to hose it down for several minutes. Water should be applied to the head and back of the animal with enough applied so it runs off the back

Sprays to cool milking stock should consist of large water droplets (Pakistan).

and down the sides. This should reduce respiration rates to 60 breaths/min. If the cow is severely stressed prior to cooling, with open-mouthed and laboured breathing and excessive salivation, she may return to the feed trough and start eating.

Sprinklers

Evaporative cooling is an efficient way of cooling cows, provided it is effective. In hot humid areas, sprinklers should always be accompanied by some form of ventilation.

Sprinklers have the disadvantage of increasing humidity and making the floor wet. Many studies show that, although they are beneficial, without forced ventilation, in the short term they do not lead to significant gains in feed intake, milk yield or expression of oestrus over a full 24 hr period. Their effectiveness depends on the humidity because the drier the air, the greater the decrease in temperature.

Sprinklers need to be suspended 2.3 m from the ground above the feed troughs with the water directed to the back of cows. The droplet size should be medium to large, depending on the humidity. It is important to install a filter at the beginning of the waterline and the sprinkler nozzles should be easily removed for cleaning. The nozzles should be directional, if possible, so that for major prevailing wind shifts, they can be adjusted to reduce wetting of feed.

Applying water to cows every 5 min reduces heat stress more so than every 10–15 min. Ideally, cows should be sprinkled for 1–3 min, applying 1–2 mm of water per 15 min cycle. The pipe size depends on the length and area of the shed to be sprinkled, the number of sprinklers and the flow rate. Farmers should use 32 mm diameter pipe for up to 30 m length or 51 mm diameter pipe for 60–150 m length. Nozzles should be spaced at twice the radius of their throw; for example, every 2.4 m for nozzles with a 1.2 m radius.

Fans and water cooling in the hot dry tropics (Pakistan).

Cows can be hosed down at the same time as their udders and teats are cleaned in preparation for milking, but this should occur at least 30 min prior to milk harvesting to minimise water contaminating the milk.

Cooling fans

Providing a cooling fan increases the rate of cooling. A simple electric fan, which can be moved around the shed to be strategically placed behind any cow, is an effective method to increase the cooling rate of heat-stressed cows after being hosed down.

Permanent ceiling fans can be arranged in many ways. A 0.5 horse power, 0.91 m diameter fan rated at 5–6 m^3/min will blow a distance of 9 m, while a 1.0 horse power, 1.21 m diameter fan rated at 9–10 m^3/min will blow a distance of 12 m. The direction of the fans should be with the prevailing wind. In wide sheds, the side-by-side spacing width of 0.9 m fans should be about 6 m, whereas 1.2 m diameter fans should be spaced 9 m apart. They should be positioned about 2–2.2 m above the floor. Fans should be tilted so they blow down to the floor directly under the next fan (about 30° from the vertical).

Exhaust fans, that do not require external power, may be worth considering to encourage greater heat removal through the roof. Household fans on stands may be used to improve heat loss from the higher yielding cows.

12.3.5 Allowing cows outside overnight

Management systems can reduce heat stress by providing access outside the shed during the cooler afternoon and evenings, say after 1700 or 1800 hr. This will promote more effective conductive heat loss than inside, hence restoration of body comfort.

Moving cows outside following the afternoon milking could be combined with a change in routine milking management on small holder dairies: namely having a specific milking area within the shed, rather than milking cows where they are tethered. Milking parlours are easier to keep clean than the entire shed and the facilities can be sterilised periodically to improve milking hygiene.

It is difficult to monitor signs of oestrus in tethered cows. Providing cows with the opportunity to mingle in outside yards will allow them to seek out any cows on heat.

12.3.6 Feeding management

The digestion of fibre produces more heat in the rumen than the digestion of other carbohydrates. Therefore, offering most of the daily forages during the cooler periods of the evening will reduce internal heat production, particularly if the forages are high in fibre, which tropical forages usually are. Rations high in rumen-degradable protein can also depress appetite during periods of heat stress. Cows consume about two-thirds of their feed during the cooler evening and night. Feeding smaller amounts more frequently will reduce the likelihood of forages drying out and losing its palatability. Feed the best quality forages at night and feed more concentrates during hot periods.

The cooler the drinking water the better for both intake and temperature balance in the body. Offer more salt to replace the minerals lost in sweat.

When given continual access to feed, cows actively seek it from 0500 to 0900 hr and again from 1700 to 1900 hr. Milkings should be finished prior to 0600 and 1800 hr. Cows prefer feeding and watering following milking, after which they should be offered a dry surface on which to rest, preferably dirt. It is important that cows stand for at least 30 min after milking so the teat end can close to protect the teat canal from bacterial invasion.

Heat stress can lead to higher incidences of lactic acidosis. Depressed feed intakes will, firstly, reduce saliva production, which buffers the rumen against rapid changes in pH, and, secondly, reduce rumen contractions, hence movement of digesta out of the rumen. Furthermore, rapid respiration rates for lengthy periods can reduce the concentration of sodium bicarbonate in the saliva, reducing its buffering capacity even further. Cows may also preferentially select concentrates and reject forages, predisposing them to acidosis.

Reduced forage intakes can decrease milk fat contents, while milk protein (or SNF) contents may fall due to lower dietary energy intakes. The immune system of heat stressed cows being under greater strain, would make them less able to cope with sub-clinical mastitis, which would become apparent in higher levels of somatic cells in the milk. Not only would heat stress reduce milk yield, but its lower milk solids and higher somatic cell counts would reduce milk returns even further through lower unit returns for the milk.

Feed manger height is best when cows are eating with their heads down, to minimise wastage and obtain highest intakes. Cows produce 17% more saliva in this position compared with feeding with their head in a horizontal or raised position.

13

Housing systems

This chapter describes the key features of well-designed dairy sheds.

The main points in this chapter

- Sheds should be located to maximise natural ventilation.
- Cement floors should be sloped for manure management and be non-slip for cows' comfort.
- There should be sufficient watering points or troughs for all stock.
- Cows can be maintained in tie stalls or be loose housed. They can be provided with free stalls or have an open lounging area.
- Additional health facilities should include treatment areas for sick stock and a calving area to permit close attention.
- Young stock should be housed according to age.
- Access to an outdoor area provides for more effective night-time cooling.
- A separate milking parlour will aid in milking hygiene.
- Good sanitation is very important for both cow cleanliness and effluent disposal. Recycling effluent will reduce necessary fertiliser inputs to forage production areas.

It is essential to provide good housing and farm layouts for rearing and milking dairy stock in the tropics to reduce the extreme effects of air temperature and humidity. Such control improves milk production by reducing stress and disease hazards and also making herd management easier. Environmental control includes sanitation and effluent management.

This chapter discusses housing for non-grazed dairy herds. Housing requirements for milking herds with access to pasture for grazing are simpler because they only require sheds for milk harvesting and for milk rearing calves. The features of shed design specifically to minimise the adverse effects of heat stress were discussed in the previous chapter. This chapter concludes with a detailed checklist for planning dairy shed design.

Good dairy sheds have many simple design features to improve cow and farmer comfort, hence cow performance and farm profitability. In hot climates, the following features are particularly important:

- sufficient yard space
- plentiful supply of drinking water

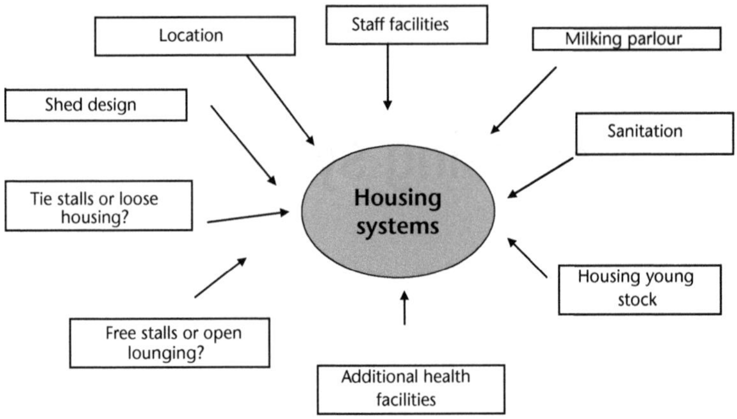

Figure 13.1. Considerations when developing housing systems for tropical dairy farms

- sufficient feeding trough space for each cow
- shade over the feeding, as well as over the loafing, area
- wire fences instead of solid walls to avoid restricting air flow
- planting trees on the western side to lower air temperatures
- sowing grass or other forages in the area surrounding the shed for stock feed
- painting buildings white to maximise reflection
- ensuring sufficient space between buildings (four times their height)
- shading over drinking troughs to keep drinking water as cool as possible
- orientating buildings to maximise winds and shade
- orientating buildings north–south to allow the sun to dry the shed
- being very wary of steep slopes.

The checklist at the end of this chapter can be used when assessing the specific requirements for shed design of particular farms.

13.1 Housing the milking herd

13.1.1 Building for the cow

It is essential that when designing housing systems, farmers keep the following key principles in mind:

- To achieve cow productivity, farmers must consider the twin issues of cow comfort and health, where health is the freedom from infections, injuries and metabolic problems.
- To achieve cow wellbeing, farmers must look at the twin issues of housing and management, where management covers all the normal farm practices associated with dairy farming.
- Addressing cow comfort and housing means that the farm's physical facilities must be optimised before putting cows in sheds.

Cows grazing in a paddock have access to forages virtually all the time and to concentrates at least twice per day, at milking. This should also be the case when cows are maintained in sheds. They should be encouraged to make at least 12 trips each day from their place of rest to the feed and water troughs. This will only occur if the cow 'feels good', has healthy feet and the route is safe and comfortable.

There are six key housing aspects of cow health and wellbeing, namely:

1. **Water**, with access to clean, palatable water at least 21 hr/day.
2. **Light**, with at least 6 hr/day of darkness.
3. **Air**, that is fresh and clean.
4. **Rest**, with a dry and comfortable place to lie down for at least 13 hr/day.
5. **Space**, so cows can walk to feed and water troughs from their free stalls without fear.
6. **Feed**, in that cows can eat a palatable and well-formulated feed for at least 21 hr/day.

13.1.2 Location

Factors to be considered when choosing land for dairy farming are fertile soils, no evidence of flooding during the wet season, convenience for transportation, proximity to milk collection centres, access to supplies for farming and, of greatest importance, year-round supplies of sufficient clean water. Distance from factories (which may release bad odours or lead to pathogenic infections) and to urban areas (where pollution from farm effluent will not be tolerated) should also be key influences on location.

13.1.3 Shed design

This has been covered in some detail in Chapter 12. The essence of good shed design is making the best use of natural ventilation, with open sides, a high and well-pitched roof together with an open vent along the top.

Concrete alleys of loose houses should slope in the direction of the movement of manure, with a 3% slope down the shed, but level from side to side. They must be easy to clean and concrete makes a smoother surface than bricks. It must allow for quick and free drainage, with all effluent directed towards a tank, either for biogas or recycling as fertiliser. In countries where dried manure is a value-added product for kitchen fuel, floor designs may differ slightly to those in other countries, where all effluent is directed to open drains.

There are several options for constructing the roof, such as iron, bricks or even bamboo/mud in a low rainfall area.

It is important to construct wide laneways to include the feeding area. By raising the walkway, building costs would be reduced and farmers will get a better view of their stock, which would be below walkway level.

13.1.4 Tie stalls or loose housing?

There are two basic types of housing: tie-up housing (tie stalls) and loose housing.

- With tie stalls, each cow is restrained in a stall. Feed is delivered in a trough in front of the cows. Milking takes place individually in the stall, by bucket or machine (bucket

Good ventilation and grass supplies are important in tropical dairy systems (Malaysia).

or pipeline milking). Manure is collected in a gutter. When cows are tied up all year round they may get feet problems and become stiff. Heat detection demands more attention. A high incidence of trampled teats can be another problem with tie stalls.

- With loose housing, cows are not tied up and can walk around freely. Such systems usually have a loafing area and a lying area, with the feeding area separated from the lying area. The cows are forced to walk frequently. The manure is spread over a large floor area so is collected by scraping the dung by hand (or using machinery) into a manure pit or a channel. With adequate water supplies, rapid flushing of large amounts of water can clean alleys. Milking is usually carried out in a milking parlour. The feed trough is separated from the loafing alley by either a feeding rack or wire rope.

Throughout Asia, dairy cows have been traditionally housed in tie stall sheds, but this is gradually changing as SHD farmers realise the benefits of providing stock with space to move around.

13.1.5 Free stalls or open lounging?

Loose housing can be of two types, either with a common lying area with open lounging or with cubicles, known as free stalls.

- In open lounging systems, cows can lie down anywhere. The floor can be earthen or cement, generally with bedding material, with the base being well drained. In dry climates, earthen floors without bedding can be used as long as the dry manure is frequently removed. The loafing area behind the feed troughs should be cement and at least 3 m wide.

- With free stalls, each cow is provided with a cubicle, which she may enter and leave at will. Cubicles can be arranged in a single row or in more than one row with a central feeding alley or with feeding alleys along the sidewalls. The cubicles can be arranged with cows facing one another (head-to-head) or the other way around (tail-to-tail). With the tail-to-tail arrangement, a central loafing alley with a width of 2.2 m between the cubicles is needed. If the cubicles are head-to-head, two loafing areas behind each row are necessary. Usually one of the loafing alleys is combined with the feeding alley behind the feed troughs. The total housing area needed in this case is no larger than the tail-to-tail arrangement.

13.1.6 Free stall layouts

The size of the cubicles depends on cow size. For cows:

- weighing 400–500 kg, they should be 104–109 cm wide and 198–208 cm long
- weighing 500–590 kg, they should be 109–114 cm wide and 208–218 cm long
- weighing 590–680 kg, they should be 114–122 cm wide and 229–244 cm long.

The surface of the free stall should have a consistent fall of 2%. It should:

- be comfortable to the cows to encourage high occupancy
- prevent hock damage and other injuries
- be easy to clean and be durable
- be cost effective to install.

The ideal lying surface is soft, absorbs moisture and does not promote the growth of bacteria. When cows are forced to lie down on hard surfaces, they do not lie down for long, are more easily unsettled and may develop knee and hock lesions and swelling. The floor of the cubicles can be earthen with a thick layer of bedding, or can be solid, with a soft top layer. The simplest bedding is packed earth or sand, this being inexpensive but requires care to maintain a flat surface. Sand is quickly pushed around by cows, so it should not be used with mechanical or liquid manure handling systems because it fills up storage tanks and is very abrasive, damaging equipment such as manure pumps.

A concrete foundation with a disposable bedding of chopped straw, sawdust, wood shavings or crushed corn cobs is more common in Europe, because rice hulls are not readily available. Rice hulls would make ideal bedding for free stalls but their high silica content could damage liquid manure handling equipment. Hard surfaces should have a slope of at least 1% so that urine will drain into the alleys.

Rubber mats and cow mattresses are new innovations for free stalls but the thin ones (10 mm or less) have short life spans. Thicker ones (25–30 mm), although more expensive, are likely to be more cost effective in the long term.

The front of the cubicles with thick bedding should be 1.2 m high and the back 0.85 m. In cubicles with solid floors, these measurements are 1.1 and 0.75 m, respectively.

To prevent the cow from soiling the cubicle, shoulder and head rails are needed. This provision will force cows backwards when they stand up. The distance of the adjustable shoulder rail to the back of the cubicle, measured diagonally, should be about 1.8 m; the height to the cubicle floor may vary between 0.9 and 1.05 m.

13.1.7 Making the choice

The choice of free stall versus open lounging is an important decision to make because the pipework in free stalls is very expensive. In virtually every feedlot shed I have seen in SE Asia, cows are provided with free stalls rather than an open lounging area. Ideally, comparisons should be made of the two housing systems in countries where labour supplies (and costs) are less of an issue. The bedding of free stalls and open lots could be rice hulls, in plentiful supply throughout Asia. Considerations when planning free stalls are summarised in Table 13.1.

Table 13.1. Considerations when planning free stall sheds

Elements	Considerations
System design	Allow for easy cow movements around the shed
	Stalls should be comfortable and clean
	Should allow for regular manure removal
	Provide easy access for machinery to feed out
	Allow for multiple routes from stalls to feeding area, to overcome dominance within herd
	Feeding space and free stalls should be proportional
	Stalls, feeding and watering spaces should be arranged in modules
Modular design	No more than 60 stalls per module
	Modules can be manipulated to match site constraints
Stall sizing	Depends on cow size (weight and body dimensions)
	Size based on largest 25% of cows in group
	Provide adequate room for lying down, as well as necessary forward and sideways lunging to stand
Stall slope	4% slope from front to rear
	3% lateral slope, so cows lie in same direction facing down slope, to minimise teat injuries
Stall curb	Minimum of 200 mm above alley way
Alley width	3–3.6 m wide avoiding sharp turns
Crossovers	Locate every 20–25 stalls, that is every 24–30 m
	4.5–4.8 m wide, so one cow can drink and two cows pass behind
	Elevate floor slightly above stall curb, allow crossover to drain into alley
Bedding material	Inorganic (sand or rubber)
	Organic (sawdust, straw, hay, composted manure or rice hulls). At least 100 mm thickness of rice hulls, sawdust shavings or sand.
	Consider hygiene, degree of compression when cow lies down, performance over time, how it will interact with effluent disposal system

Open lounges create their own problems of regularly removing and cleaning the bedding and ensuring all cows will use it in preference to lying on dirty cement walkways, and hence increasing potential mastitis-producing areas.

13.1.8 Considerations in flooring

The required floor slope depends on the cleaning method adopted and the natural slope of the site. A minimum slope of 0.5% (1 in 200) is required to prevent pooling of water. Steep cow alleys (greater than 1 in 15) will result in cows not standing perpendicular to the feed trough, but standing uphill. Flood washing of the cow alley ideally requires a slope on between 2.5% (1 in 40) and 3.0% (1 in 30), to achieve acceptable wash with minimum amounts of water. If the slope is less than 2.5%, the volume of flood wash water required increases to get extra water depth to compensate for the lack of slope. There should not be any sideways slope on cow alleys with flood washing. If flood washing is not used, the cow alley should have a 2.5% slope away from the feed trough. Hose washing and dry scraping are not slope sensitive, so the site slope will influence the alley slopes. The feed alley needs a concave floor (with 50 mm fall) sloping away from the feed to the centre of the alley to prevent run-off water mixing with the feed. The long-ways slope should be the same as the cow alleys.

Cows can be standing or walking on shed floors for up to 10 hr/day. In tie stalls, cows should be provided with rubber mats to cushion the hard concrete floors. With loose housing, concrete floors should be roughened to reduce slippage, but not too rough to wear out hooves, leading to lameness problems. Another alternative is to install grooves or patterns in new concrete floors. These can even be cut into old concrete floors.

Grooves should be positioned so that at least one of the four feet lands in a groove whenever cows put their feet down to stop slippage. Parallel grooves should be spaces about 15 cm apart, 12 mm wide and 12 mm deep. Although placing grooves perpendicular to the length of the alley will maximise their effectiveness, they may compromise removal of shed effluent, so they should be located parallel to normal alley-way cow flow.

Anti-slip aggregates mixed in with the top layer of concrete or epoxy floor coatings can be applied to existing floors. Another alternative in areas of high cow usage, such as in feeding alleys, is construction of synthetic rubber floors such as those used on athletic tracks.

13.2 Feeding the milking herd

13.2.1 Water troughs

Stock must be provided with sufficient drinking water at all times and the system should be able to supply at least 20 L/cow/hr to meet likely peak demand. The optimum temperature for drinking water is 15–17°C. Water troughs should be well separated (but within 15 m) from feed troughs, with the same water trough servicing adjacent pens. Trough design should allow for regular, easy cleaning with a removable bung for complete drainage.

Each cow should be provided with 75 mm of linear watering space in free stall sheds, while for circular water tanks, one watering space (60 cm of tank perimeter) should be available for every 15 to 20 cows. A water depth of 15–20 cm helps keep water cooler, fresh and easier to clean because less debris accumulates. The optimal trough height is 60–90 cm from the ground to the top of the trough.

If cows are tethered, it is possible to install individual cattle drinkers, with two animals sharing one drinker. To ensure water is freely provided, troughs need to have float valves, rather than depend on a hose whenever the farmer decides to refill it. A single watering trough system can control water levels in up to 10 small troughs just by one float valve in a central reservoir with syphon pipes leading to all the troughs.

13.2.2 Feed troughs and feeding strips

It is important that all stock can eat comfortably with minimal competition. Whether feeding into troughs or onto cement feeding strips, 70 cm of feeding space should be allocated per mature cow. For a feeding strip or trough when cows eat from both sides of the strip, hence face each other to eat, this equates to a total feeding strip allocation of 35 cm/cow. Trough space can be reduced to 45 cm/head for 6-month-old cattle, or to 55–60 cm/head for stock 18 months old.

With feed troughs, the width should be related to the reach of the animals, which is up to 80 cm for mature dairy cows eating from the base of the trough 30 cm above ground level. The base of the trough should then be raised 10–30 cm, with the front 50 cm above this. Cows in tie stalls are likely to have restricted reach.

Feeding strips should be positioned 7–15 cm above cow feet level, with a nib wall 40–55 cm above the feeding strip, to protect the feed from the effluent. Feed barrier walls wider than 15 cm can restrict how far stock can reach into troughs or onto feed strips.

13.2.3 Feed storage

Feeds should be stored in secure and vermin-proof locations with easy access for the stock. They should be located well away from waterways. Most feeds are sourced in bags, although larger farms may handle feeds, such as by-products, in bulk. In these cases, good road access for trucks is essential, with the feed bunkers under cover. Such bunkers should have concrete floors and brick, concrete block or prefabricated or reinforced cement walls. For storing by-products with very low dry matter contents, soil ramps adjacent to these bunkers will aid discharge from trucks.

Forages can be conserved dry as hay or straw, or moist as silage. Hay or straw must be stored in weatherproof buildings and should not be stored in confined places if it is likely to be moist, where it can catch on fire. Silage must be continually covered with plastic sheets to exclude any air prior to feeding out. Silages can be stored above ground in bunkers or below ground in pits. In-ground pits are best for long-term storage but above-ground storage is preferred in areas with high rainfall. Buns or stacks are silage units without structural support, but with their high surface area-to-volume ratios, plastic costs per tonne silage and wastage rates can be high. Bunkers or clamp silos are the most common storage systems, while individually covered bales or stretchable bags are often used in regions well serviced by silage making contractors. The dimensions, particularly their width, should be determined prior to their construction, based on the predicted daily rate of removal from the silage stack.

Silages can produce a lot of liquid, called leachates, during storage, which is very corrosive and lethal to aquatic life if it escapes into rivers or streams. The production of leachates can be reduced by incorporating dry material, such as rice straw or rice bran, at the base of the silage bunker.

13.2.4 Preparing concentrate mixtures

It is rare for milking cows to be offered only one type of concentrate, because of their high nutritive demands and the poor nutritive value of tropical forages. Concentrate mixtures usually contain a variety of energy and protein supplements together with minerals, usually macrominerals and, for high-yielding cows, maybe even microminerals.

One very simple method of hand mixing a complex concentrate mix for storage or immediate feeding is to spread each ingredient on top of each other in layers (even the mineral pre-mixes) then collect the mixture into bags or buckets by shovelling it perpendicular to the floor. This method is ideally suited for dry concentrate ingredients, although it could be used each day for offering mixtures containing wet by-products, such as cassava or soybean waste.

Molasses blocks are a good supplement and can be used as carriers of non-protein N or minerals. The blocks are solid, and fairly easy to make, transport and store. Molasses intakes should not exceed 20% of the total intake, because at higher levels it will depress digestibility.

If the basic ingredients of concentrate mixtures are stored, a space must be allocated to blend them into a formulation. Hand mixing may be practical with small herds of stock, but a mechanical mixer is preferred to ensure complete blending of mixes for larger groups of animals. These formulations are either placed on top of forages or fed out following removal of residual forages from the feeding troughs. With very large herds, however, a mixer wagon is a good investment to ensure consistency of ration formulations. Such mixtures are called total mixed rations, or partial mixed rations if stock also have access to grazed pasture.

The tradition of mixing water with the concentrate portion of the ration is widespread throughout Asia. Farmers wrongly believe that cows will better use concentrate slurries compared with dry mixes. This may be one way of providing additional water, in housing systems where there is limited free access and stock are only provided with restricted times to drink. However, given free access to water, there is no need to feed slurries of concentrates. Cows can be fed their roughage portion on a flat area of concrete with no sides. It is also easier to clean such a flat feeding area. Therefore feed troughs could become superfluous.

13.3 Additional features of dairy shed design

13.3.1 Additional health facilities

To ensure good health care, sufficient health facilities are needed. Sheds should include additional stalls for such purposes. These are as follows:

- A treatment area for confining animals in heat, artificial insemination, routine health checks, pregnancy diagnosis and examining sick cows. Because animals are usually separated when they leave the milking parlour, the treatment area should be located close by. The width should be at least 0.7 m per cow and the length 3 m. It is convenient to have a separate movable veterinary drug and equipment box for treating hoof problems, for trimming hoofs, taking blood samples, and so on. There should be one treatment stall per 20 stalls with a minimum of two.
- A separation area to treat sick cows properly and prevent the spread of disease. It should be located close to the milking parlour. The cows can be housed in this area in 1.2 m stalls, which could be changed into pens. Drinking water should be available, with concrete floor and gutters to allow for frequent cleaning and sanitising. There should be one treatment stall per 30 stalls with a minimum of two. With tie stalls, there is little need for a special separation area.
- A calving area, to permit proper attention at this critical time. With loose housing, cows may need to be tied up in stalls and should calve down away from the milking herd and close by the calf pens. As with the separation area, ease and thoroughness of cleaning and sanitation are key features. There should be one calving stall per 30 stalls with a minimum of two.
- Foot baths should be available; they should be at least 2 m long and 0.15 m deep. The width should be the same as the passage to prevent cows from bypassing the bath without using it.

13.3.2 Housing for young stock

The area needed to house young stock depends on:

- their number
- the type of housing
- the calving pattern of milking herd
- the replacement percentages of the milking herd
- the division of the herd into age groups.

A good management practice is to divide young stock into the following four age groups:

1. 0- to 1-month-old milk-fed calves
2. 1- to 5-month-old milk-fed and weaned calves
3. 5- to 22-month-old weaned heifers and yearlings
4. 22 months and older pregnant heifers.

Group 1 should be housed in clean and disinfected individual pens (0.6 × 1.3 m), separated from the cows. The pen partitions should be solid and smooth, 1.5 m in height. The floor should consist of wooden slats 20 mm apart.

Group 2 can be housed in tie stalls (0.7 × 1.5 m) or group pens. Group 3 can be housed in tie stalls (0.8–1.0 m wide × 1.7–2.1 m long, depending on live weight), group pens or free stalls. Group 4 can be housed with dry cows, although they will have different feed requirements.

It is easier to observe calves, and there is less risk of disease transfer, when they are individually penned. Ideally, the pens should be moveable and be enclosed with three draught-proof walls. Each pen should have 1–2 m^2 of floor space with a well-drained floor and, if desired, clean bedding each day. An individual pen 1.5 m × 0.8 m is the minimum required up to 4 weeks of age while a pen 1.8 m × 1.0 m will suffice for an 8-week-old calf.

13.3.3 Calf cages

Containing each calf in a raised metal cage provides the best housing because they are isolated from each other, live in a well-ventilated and clean pen, are easier to feed and water (because the rearer does not have to bend down to feed them) and, of most importance, allows for much easier individual surveillance. Calves should live in cages up to 3 weeks of age, after which they can be moved to individual pens.

Each cage should be 110 cm long, 75 cm wide and 105 cm high (including the legs). The rear of the cage is a removable gate to put the calf into or remove it from the cage. The floor can be made from wood or plastic, with the floor lattice perpendicular to the length of the cage. The space between floor boards should be 1.5–2 cm. The cage should stand on legs and be 30 cm above floor level. To avoid injuries, there should be no sharp edges inside the cage. The metal bars should be about 15 cm apart: any narrower can lead to injuries if calves get their legs stuck between the metal bars.

13.3.4 Outdoor area for night-time

Once the sun sets, outdoor areas are cooler than inside any shed. Saudi Arabian dairy feedlots have them, as do many Australian dairy feedlots. They are much less common in Asia because of the high rainfall and problems with mud. Where cows are commonly housed out of doors, they should be an integral part of any shed design. Whether the floor needs to be concrete or brick or just plain dirt is another issue.

Cows will eat more during the cooler part of the day, so provision must be made for adequate feeding space. It is better to provide the bulk of the forage in the evening, particularly during summer, so feeding space is important. Free access to adequate watering space is essential. Also, because it is easier to observe oestrus when cows are loose in a yard at night, permanent lighting is important.

A lounging area for the milking herd should:

- be located in a well-ventilated area
- be well drained
- have a dirt or gravel surface
- provide at least 5 –10 m^2/cow
- incorporate water troughs
- incorporate troughs for feeding forages.

13.3.5 Stock-handling facilities

It is important that stock can be handled safely when individually separated from the herd. The farm should have a set of cattle yards that includes a race for AI, a crush for close veterinary attention and treatment and, if desired, a set of scales to weigh stock routinely.

13.3.6 Storage of veterinary drugs and other dairy equipment

Veterinary drugs should be stored in a dark cupboard or, if required, inside a refrigerator. For security purposes, the cupboard should be lockable. Other dairy equipment, such as detergents and sanitisers for cleaning milking machines and equipment for trimming feet or dehorning calves, should also be stored in a specific location. Milk-harvesting equipment, such as buckets, sieves and milk cans should be stored off the ground and hanging upside down to aid drying.

13.3.7 Bull pen and mating yard

Although many farmers use AI to mate their cows, herd bulls are often kept to mate heifers or cows that have been unsuccessfully inseminated several times. Bulls should be kept separate from the adult cow herd except when used for mating. A mating yard will allow for cows on heat to be separated from other stock for better control of the bull's activities.

13.3.8 Farm office

Maintaining good farm records is much easier in a farm office. An area at home or in the dairy shed should be dedicated to keeping records. It must have a desk and good lighting. It must be a quiet place to set up the office files (preferably in a filing cabinet), computer and office supplies. The how and when of keeping farm records depends on the person recording them. Computers are very convenient, but require money to purchase and skills to operate efficiently. Record keeping should be given as high a priority as other farming activities.

Suitable chairs and tables should be included for business meetings with service providers and other farm-related visitors.

13.3.9 Staff quarters

Farm staff should be provided with space to eat and relax when off duty. This could include a shower and toilet, food preparation area and storage for their work clothes.

13.4 Milk harvesting

Once a decision is made that cows are allowed outside to cool down at night, it seems logical to plan their movements around milking times. A separate milking area has many advantages. It can be centrally located, requiring only one supply of hot water (for cleaning animals and equipment). Daily hosing down and weekly disinfecting will minimise environmental constraints to good milk quality. Furthermore, if farmers eventually decide to invest in machines for milking, one milking area minimises problems with low vacuum pressures in long air lines.

Milking hygiene is very poorly managed in many countries, particularly those where farmers are not rewarded for their milk quality. However, it will become important as

A well-managed and well-constructed dairy shed in Indonesia.

milk processors invest in the industry and when farmers consider value adding with other dairy products, such as cheese, that require high-quality raw milk.

Cows can be milked in a variety of ways. Hand milking is very common throughout Asia, while bucket milking is used for cows in tie stalls. Loose housed cows are usually milked in a separate milking area, which can be specifically designed for ease of cleaning and occasionally sterilising. Hot water is essential for thorough, twice-daily cleaning of all milk-harvesting equipment. Ideally, milk-cooling equipment should be installed, unless there is close access to a milk collection centre.

Bucket milking is the simplest and least expensive system to install, but the milk must be carried to the cooler. This type of system is usually used in small herds and where cows are milked on a level floor of the shed or milking area. A separate vacuum line can be built around the entire shed if the cows are to be milked in tie stalls. The labour of carrying milk to the cooler can be avoided by installing a transfer system, which consists of a small receiving tank (say 30 L), including a built-in filter, mounted on wheels so it can be moved around the shed. Pipeline milkers transport the milk directly from the cow's udder to the milk cooler and are usually installed in milking parlours where the milker stands below the level of the cows. Although they are expensive, they save backbreaking labour and are usually designed to be cleaned in place: a feature that not only saves labour but helps to ensure good sanitation. Details of maintaining the milking equipment are discussed in section 14.4.

13.4.1 Designs of milking parlours

All parlours should have a high-quality concrete floor and metal rails for durability and ease of cleaning. Walls are not required, but, if included they should be at least plastered masonry walls. The pit where the milker stands should have a floor level 90 cm below that of the cows for the most comfortable working position. The number of stands is determined by the allowable milking time of the herd or time taken by the cows to eat the concentrate rations usually fed. Parlours can be of the following types:

- **Abreast parlour.** The milker and the cows share the same floor space. Cows can enter and leave individually. The stands should be 1–1.1 m wide when a bucket milker is used or when the cows are hand milked, while 0.7–0.8 m is adequate when a pipeline-milking system is used. The width for the milker should be 0.6–0.8 m. The main drawback is the relatively long distance to walk between milking points, and the cows obstructing the milker.
- **Tandem parlour.** This allows for individual care of the cows, but has larger space requirements.
- **Walk through parlour.** Cows enter and leave in batches. The narrower width can be an advantage if fitted into an existing building.
- **Herringbone parlour.** This provides a compact working area and allows feeders to be fixed to the side walls. Three or four stands on each side should suffice for up to 80 milking cows. Its popularity is due to its simplicity and high capacity (measured as cows milked per man hour). However, the risk of cows kicking the milker is greater than in parlours where the milker stands beside the cow.
- **Rotary parlour.** Because these are expensive to install, they are better suited to much larger herds.

It is best to equip milking parlours with grain feeders that allow each cow to be fed in proportion to her milk production. Because cows expect to be fed while being milked, they will enter the parlour more readily, thus saving some labour. Each cow can be manually fed her ration using a measuring scoop.

An entrance to a milking parlour that is straight, with no turns, will ensure a smooth and convenient operation. Once trained, stock will walk readily into the parlour. A single 1 cm step will help keep manure from being carried into the parlour. An exit leading to an uncrowded area will facilitate the flow of cows. It should be narrow (70–80 cm depending on the size of the cows) to keep the cows from turning around.

13.4.2 Milk room and cooler

Sanitation is the primary consideration in handling milk, whether from a hand- or machine-milked cow. Adequate supplies of potable water are essential for cleaning the milking equipment immediately after use. Hot water (85°C) mixed with a chemical detergent are required for effective cleaning and cold water used for rinsing. Milk should be handled in a separate area that can easily be cleaned and is free of insects, birds, rodents and dust.

The milk room should be well ventilated, with a concrete floor sloped 20 mm/m to a drain and with masonry walls having a smooth, water-resistant surface that can be easily and thoroughly cleaned. On dairy farms of sufficient size, the milk is cooled by cold water circulated between an evaporative water cooler and a milk cooler (plate heat exchange), through which the milk is passed until it is adequately cooled. Where milk is transported in cans, cooling can be accomplished through a coil, which is immersed in the can. The larger scale farmer, having a pipeline milking system, and milk collection by a tanker, will require a refrigerated cooler and holding tank.

With machine milking, the vacuum pump and engine that powers it, should be put in a separate engine room.

13.5 Sanitation and effluent management

Dairy farm effluent is both a liability and an asset to the small holder farmer. It is a liability so far as contaminating feeds, encouraging flies and animal health issues are concerned, whereas its role in fertilising forages is an asset. Unfortunately, its full potential is not realised unless all the nutrients contained in the urine are returned to the pastures. The nitrogen in urine is easily lost through volatilisation, unless it is stored and recycled to the pastures in liquid form.

Each day each cow produces 30–35 kg faeces and 12–16 kg urine plus shed effluent from washing down floors. Each day a 50 cow shed then produces 1.6–1.8 t faeces plus 700–1000 L urine, which is equivalent to 1.5–2 m^3 volume, excluding rainfall run-off. Therefore it is important to:

- design floors for efficient effluent run-off
- ensure that there are suitable gutters and sufficient wash down water to keep floors clean
- ensure that the effluent pit and pond has sufficient capacity
- ensure that the base of the pond is 1 m above the water table
- provide equipment to handle solid effluent for recycling onto the forage area
- provide equipment to remove liquid effluent, such as a pump and pipes that lead directly to the forage area
- be aware that 'social standards' are always changing
- consider an area to dry manure for bricks or cooking fuel.

13.5.1 Effluent as a liability

The management of animal manure represents a major health hazard on small holder farms, with the problem increasing with herd size, unless specific facilities are constructed. The problem is associated with a number of issues, such as:

- the quantity and quality of faeces and urine produced
- the adequacy and frequency of removal
- storage in proximity to the shed
- labour availability
- the methods of storage and disposal
- the value and use of manure
- community concern about pollution (smell, as well as contamination of groundwater and water courses).

The human health hazards are becoming more serious than previously realised, due to inadequate supervisory and sanitary measures. For example, in Thailand, a survey noted waste water from SHD farms constituted a considerable risk to public health because of its very high content of inorganic pollutants and the presence of coliform organisms (Chantalakhana *et al.* 1999). Furthermore, there was considerable contamination of groundwater supplies in nearby towns via waste water from local farms as well as via leaching from stored manure.

Not only is effluent management an environmental constraint to small holder dairying because of its potentially adverse influence on cow performance and cost of forage production, but, of equal importance, it can alter the livelihood of farmers themselves through community legislation limiting locations of farms. This can be particularly important for dairy farmers situated close to large towns and cities. In developed countries, there is now considerable effort to ensure farming activities are seen as 'clean and green' and this concept will no doubt extend into the humid tropics in years to come.

13.5.2 Effluent disposal systems

Floors should be made from concrete or easily washable material. They should be designed for efficient drainage with a good slope with wide channels for easy urine and faeces removal. A manure pit should be dug, large enough to hold the shed's manure produced over 2 or 3 days. A channel should be dug leading from the walking area to the pit and lined with concrete. The pit should be covered with a plastic sheet or banana leaves to reduce the sunlight, because this volatilises the nitrogen in the manure, reducing its value as a fertiliser. A fence around the pit will minimise risks to children and wandering stock.

It is important to collect as much urine as possible from the shed. It is advisable to use minimum water to initially wash out the manure drains, then direct the main washings from the floor into another pit or directly out to the forage producing area. An alternative effluent system is to direct all manure and shed washings into the one pit equipped with a manure pump to direct the effluent to the forage producing area, along gutters between the rows of plants.

13.6 A checklist for planning dairy sheds

Physical aspects:

- aspect of shed with respect of wind and sun
- flooring (material, slope)
- rubber mats for floors
- wooden or metal uprights
- roof height
- roofing material ('colourbond' or tiles)
- maximum natural ventilation
- feed troughs (cement, wood or other material)
- water drinkers (automatic, trough plus float valve, trough plus hose)
- location near feed-growing areas (for effluent recycling)
- design shed to allow easy stock observation
- wide aisles for ease of carrying heavy loads.

Services:

- electricity (single or three phase)
- water (drinking, cooling, washing, sanitising)
- water heating (liquid petroleum gas or fire)
- water for washing down, taps and hose for floor flushing.

Environment:

- insulation for under roof
- sprinklers for cooling
- hose for washing down hot cows
- cooling fans (large overhead or small moveable)
- fans for all milk cow areas or just for high-yielding cows
- computerised system to initiate specific cooling cycles
- plant trees, grass around sheds.

Effluent disposal:

- suitable gutters to keep floors clean
- effluent pit (suitable capacity)
- one large pit for several sheds
- liquid effluent removal (pump and pipes)
- solid effluent removal
- location of fodder growing areas for recycling
- area to dry manure and make bricks/cooking fuel.

Feeding facilities:

- separate feed-storage areas
- bird, vermin and insect proof
- silage pits
- bunkers for urea-treated straws
- fresh or dry forage feeding
- hay shed for bulk storage of purchased hay/straw
- concentrate storage (bulk or bag)
- chopper for processing roughages
- feed mixer (horizontal or vertical) for concentrate formulations
- stationary mixer wagon to process total mixed rations
- farm utility for transporting feed, stock and milk
- farm tractor for handling bulk feed
- floor space for hand-mixing concentrates.

Milk harvesting:

- hand or machine milking
- separate milk-harvesting area
- mobile mini milkers or milking plant

- vacuum lines above cows
- engine room for milk plant
- specific buckets/milk cans for each shed
- bulk milk handling equipment (vat, milk can)
- refrigerated milk vat in each shed or for entire complex
- other milk processing equipment (pasteurising, packaging)
- specific equipment for cleaning/sanitising milk handling equipment.

Adult cows:

- tie stalls or free stalls
- adequate feeding space per cow
- access to drinking water
- level of milk production
- group cows by stage of lactation
- halter and rope or neck yoke
- separate large yard for heat detection and night cooling
- handling yards and crush for veterinary attention
- AI or natural mating yard
- mating yard for bulls.

Young stock:

- separate milk-rearing shed
- separate weaned young stock shed
- suitable capacity for 25–30% replacement rate in same shed
- design of calf-rearing pens (above ground or deep litter)
- location of milk-fed calves (away from adult stock)
- material for calf pens (bamboo, wood, metal)
- area for mixing milk replacers and calf concentrates
- hot water for cleaning equipment
- feed scales for weighing all feeds
- livestock scales for weighing stock
- hospital/isolation pen for infectious diseases.

Personnel:

- office and suitable furniture
- computer facilities
- 'smoko' or tea room
- living quarters (if overnight work is required)
- lockable storage of veterinary drugs and valuable equipment.

14

Milk harvesting and hygiene

This chapter discusses the key principles of good milk-harvesting practices to ensure high milk quality.

The main points in this chapter

- Milk composition and milk quality are often incorrectly used to mean the same thing; however, composition is the content of milk components while milk quality is a measure of the various contaminants in the raw milk.
- The various measures of milk composition and milk quality are described.
- The basics of good cleaning of milking equipment can be summarised as WATCH, namely: water, action, time, chemicals and heat.
- Good milk hygiene practices make it possible to produce clean, safe milk and dairy products with less than the ideal equipment and facilities that are generally found on small holder farms in tropical environments.
- Hygiene can be split into various milk-harvesting practices, such as health and personal hygiene, environmental hygiene, milking procedures and milk handling, and finally post-milking procedures.
- When using machines for milk harvesting, a good testing and maintenance program is essential to ensure quality milk and minimal udder diseases.

Milk is the most perishable of all farm produce. Unlike other animal products, such as meat, milk is frequently harvested in very unhygienic conditions, where all too frequently, the current practices of cleaning and sterilising the containers used for its collection and transportation leave much to be desired. Not only is bacterial contamination in buckets and milk cans a major problem, but, because the tropical environment encourages rapid growth of these bacteria, the prolonged time delays in cooling the milk to 4°C, reduce its quality even further.

The government legislation controlling milk-harvesting systems in countries with developed dairy industries ensure a consistently superior milk quality that is just not currently possible in most of the SHD farms throughout Asia. Good milk hygiene

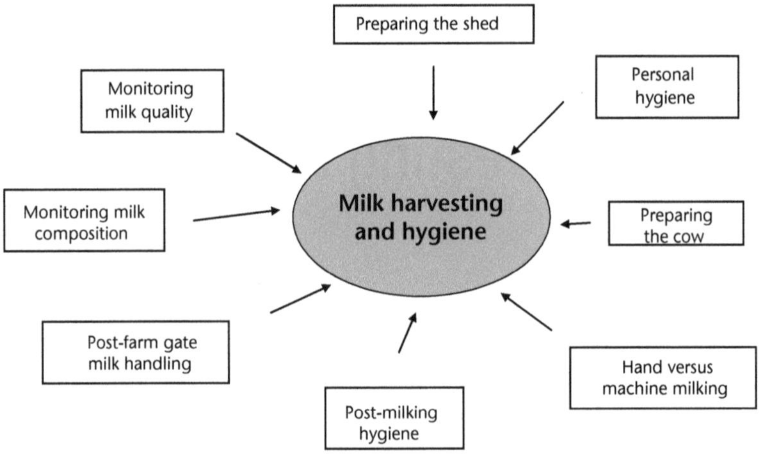

Figure 14.1. The key elements of good milking hygiene

practices make it possible to produce clean, safe milk and dairy products with less than ideal equipment and facilities that are generally found on small holder farms in tropical environments.

Good milk hygiene produces dairy products that are safe for human consumption, and that have good keeping quality. On the other hand, poor milk hygiene leads to spoiled products, product recalls (hence adverse publicity), food-borne diseases and

This milk bucket is not suitable for hygienic milk harvesting (Indonesia).

unsatisfactory or declining product image. This all leads to reduced consumer confidence in the integrity of the dairy value chain.

This chapter discusses milk harvesting, either using hand or machine milking, and the key aspects to ensure the integrity of the raw milk is not compromised by poor management and milk-harvesting practices.

14.1 Measures of milk composition and quality

14.1.1 Milk composition

This refers to the levels of total solids, milk fat and solids-non-fat or SNF (which include protein, lactose and minerals) in the milk. Milk hygiene, on the other hand, refers to the levels of various contaminants in milk, whether bacterial, chemical or any other adulterants that are detected. In many Asian countries, the term milk quality covers both composition and hygiene, whereas in Australia, it refers specifically to milk hygiene.

The **total solids (TS)** content in raw milk is estimated from its specific gravity using a lactometer, after adjusting for milk temperature. The results are not very accurate because the overall specific gravity is due to water, material less dense than water (milk fat) and material more dense than water (SNF). Therefore, milk with high fat and SNF content can have the same specific gravity as milk with low fat and SNF contents. Furthermore, milk temperature is sometimes not measured, meaning that cooler milk will have a higher specific gravity, hence a higher estimated TS content. If farmers leave their milk to cool before bringing it to a milk collection centre, it may have a higher estimated (and non-temperature-adjusted) TS content, but will also have higher bacterial contamination. The alcohol test can indicate this, thus allowing the milk to be rejected for poor quality.

Milk fat has traditionally been measured in the laboratory using a centrifuge to separate out the milk fat, which had been coagulated with a strong acid solution, although more modern techniques allow for rapid measurements. The **solids-not-fat** content is calculated as the difference between TS and milk fat. The **milk lactose** content is very consistent across all breeds and feeding management.

14.1.2 Milk quality

There are various ways and methods of monitoring milk quality, with the key ones described briefly below. Many of the methods are relatively inexpensive, with little needed in the way of equipment. Other methods may require more expensive equipment, but are more accurate and quicker.

The **total plate count (TPC)** measures the level of bacterial contamination that occurs after milking until the arrival at the milk collection centre or milk processors. It is one of the most common milk hygiene tests performed in the world. A small sample of the milk is taken and is placed with an agar growth medium. This mixture is then incubated for several days and the number of bacterial colonies that grow are counted, hence the name plate count. The greater the level of contamination, the poorer the keeping quality of the milk, and the less the farmer is paid for the milk.

The **methylene blue reduction test (MBRT)** is also a very common test for milk quality and is relatively easy to perform. It is similar to the TPC because it also measures the level of bacterial contamination in the milk. The initial mix of methylene blue and milk is blue in colour. As the level of oxygen diminishes in the milk due to bacterial activity, the blue colour changes to white. The time for this colour change indicates the level of bacterial contamination present in the milk sample. The shorter the time for milk to change colour, the more contaminated the milk.

The **resazurin test** is a 10-minute test to assess bacterial contamination through colour change, which is not as accurate as the MBRT, but is easier and quicker. The resazurin imparts a blue colour to the milk, which changes to pink and then to white over time in a hot water bath. After 10 min, the colour is compared with standards to assess milk quality.

Measuring **milk pH** helps to monitor whether compounds have been added to the milk to alter the specify gravity (and so falsely indicate higher levels of milk solids) or alcohol test, or mask the addition of water added to the milk to increase its volume. The pH should be 6.6–6.8. Lower values generally mean an acidification process due to bacterial growth, while higher values can indicate the presence of mastitis.

The **senses test (organoleptic)** usually involves trained operators tasting, smelling and visually assessing the milk consignment. It does not require any expensive equipment to perform, but is a subjective test. Any off-flavours, taints, smells and visible contaminants can be detected by a trained assessor.

The **milk temperature** is recorded at the milk processors from each tanker as it delivers the milk to monitor the degree of warming since it was first cooled, to 4°C, at the cooling plant.

In the **alcohol test**, milk is mixed with an equal volume of 68% ethanol and if it does not form floccules the milk is normal. If abnormal, the type of flocculent can indicate the type of contamination. Milk that coagulates is likely to contain high levels of acid, calcium or magnesium contaminants.

In the **clot-on-boiling test,** milk is boiled for 5 min to assess its heat stability. If it clots, it is likely to coagulate during processing (such as pasteurisation). This test measures the same characteristics as the alcohol test, but does not require the use of chemicals.

The **specific gravity** is measured using a lactometer to monitor both the milk composition (fat and SNF) and water adulteration. The density of milk is 1.028–1.034 at 15–20 °C and values below 1.028 usually indicate the presence of added water.

When raw milk is delivered to milk collection centres, it is routinely weighed then subjected to various 'platform tests' for contaminants, using the senses and alcohol tests. It can then be sampled and tested for specific gravity and fat content to monitor solids-not-fat content. MBRT and TPC (and resazurin) tests are generally measured in laboratories, rather than on the platform.

The major influence of milk composition is nutrition, breed and stage of lactation. Milk quality is the result of both intrinsic factors, such as mastitis infections, and extrinsic factors, such as :

- **the animal shed**, which should be clean, with a good supply of clean water and hot water

- **the animals**, which should be clean and have dried teats, and be checked for sub-clinical mastitis
- **the milker**, who should be healthy, wear clean attire, not smoke while milking, and use the correct milking techniques
- **the milking equipment**, which should comprise clean and sterile metal buckets that are stored upside down
- **milk storage and transport**; milk should be filtered before bulking and delivered rapidly to the milk collection centre in covered containers.

14.2 The science of milk quality

14.2.1 Factors influencing the growth of bacteria

It is important to understand the principles, firstly, of bacterial growth and, secondly, of cleaning and sanitising dairy equipment. This allows farmers to more competently and confidently consider local issues and find solutions to poor milk quality.

Bacteria have four basic requisites for growth, namely:

1. **Food.** Bacteria thrive in conditions supplying food and moisture, such as milk. Therefore, poorly cleaned utensils are ideal for bacterial growth.
2. **Warmth.** Although each species have optimal temperatures, they generally grow very well within the 20–40°C range.
3. **Moisture.** Most foods contain sufficient moisture for bacterial growth. Clean utensils should be stored upside down to facilitate drying.
4. **Time.** If provided with food, water and 37°C, bacteria will reproduce every 20 min. Each bacteria then can produce 500 new bacteria within 3 hr and 1 000 000 within 10 hr.

On farms where milk is not cooled until 1 or 2 hr post-milking, TPC levels can quickly become very large. Even when stored at 4°C, TPC levels increase, particularly if not delivered to milk processors within 1 or 2 days.

14.2.2 The principles of cleaning and sanitising dairy equipment

The cleaning and sanitising of milk-harvesting equipment are two separate and distinct operations that must go together to ensure minimal bacterial contamination. Put simply, cleaning removes residual milk from surfaces, whereas sanitising removes bacteria from cleaned surfaces. Basically, detergents are chemical removers of surface deposits. The general guide to cleaning is to:

- dry clean and remove all loose dirt and debris
- rinse or wet the surface, using cold or warm (not hot) water
- hot wash using a detergent solution that holds contaminants (or soils) in suspension for a short time
- rinse with cold water and drain
- apply sanitiser to contact surfaces and allow to dry.

The basics of good cleaning and sanitising dairy utensils can be summarised by the acronym WATCH, as follows:

- Water. Water quality, such as the level of sediment or bacteria, is very important.
- Action. This means using mechanical action (with pumps or vacuum) to encourage agitation, or manual cleaning, such as using a brush.
- Time. It should be long enough for the chemical to work but not too long for redepositing of soils. The longer a surface is manually cleaned, the better. Cleaning in place (CIP) systems are the best because they require the least physical effort.
- Chemicals. The chemicals should be matched for the job, used at the recommended dilution rate and in the right sequence. Protective clothing should be worn if necessary.
- Heat. Chemical activity doubles for every 10°C over 50°C. Excess heat can denature some sanitisers.

Different types of detergents have different roles. Neutral detergents are the easiest to use because they require no skin protection. Alkali detergents remove protein, fats and carbohydrates, whereas acid detergents are best at removing milk stones and hard water scale. Good cleaning practices require regular use of both alkali and acid detergents. Their effectiveness is compromised when used in lower temperature water.

Milk stones are hardened deposits formed from residual milk, bonding to metal, rubber and plastic surfaces. Water high in dissolved minerals (hard water) will form a hard water scale. Both these residues provide a suitable environment for bacteria, which are released into any milk coming in contact with them. Milk stones will become more of an issue as hot water becomes more widely used.

To provide some guidelines on optimum water temperatures, Australian dairy cleaning standards are as follows:

- pre-rinse: not exceeding 50°C
- flush cleaning: 80°C
- acid cleaning: 80°C minimum
- sterilising: 90°C minimum.

14.3 Management practices for good milking hygiene

There are 20 key factors for small holders to produce clean milk. These are presented in Table 14.1, together with the associated on-farm practices.

These management factors can be split up into various milk-harvesting practices such as:

1. health and personal hygiene
2. environmental hygiene
3. milking procedures and milk handling
4. post-milking.

Table 14.1. Key factors and on-farm practices to undertake to produce clean milk on small holder farms

Key factors to produce clean milk	On-farm practices
Prepare the shed	Repair any holes in floor, clean the floor and wash with disinfectant
Personal hygiene	Use clean clothes, carefully wash hands; don't milk if you are sick
Prepare for milking	Have ready the udder cloths, buckets, stool, basket (for dirty udder cloths), strip cup, muslin cloth and milk can to save time; place milk can outside shed
Pre-milking cleanliness	Do not use milk containers for any other purpose, all equipment must be clean, sanitised and dry
Cow comfort	Gentle handling of cows, maybe offer some concentrate, but not roughage
Cow cleanliness	Brush cow to remove dust, wash udder and teats, and dry teats
Cow disinfection	Use one cleaning cloth per cow soaked in hypochlorite (1 teaspoon/5 L water)
Reduce disease transfer	Use one cloth per cow; put used udder cloths into separate basket; don't let milk drip/spill onto floor
Water quality	Only use good-quality water for washing cows and containers
Pre-milk each teat	Strip milk each teat into cup to check for mastitis and remove initial milk
Hand milking	Use fast steady speed, use 'hand squeeze' not 'hand strip' technique; don't use oil, water, milk or spittle as lubricant; use hand cream if necessary
Machine milking	Routinely replace rubber linings, sanitise after use; follow correct maintenance schedule; open tops of milk cans in cooling unit to facilitate heat dissipation
Timeliness of milking	Start milking within 30 seconds of washing udder, cow's let-down lasts 5–7 min
Teat dip	Dip each teat into iodine solution; can use all in cup if solution is still clean
Bulking milk	Quickly strain into milk can through muslin to remove contaminants; put lid on can
Cooling milk	Take milk to a milk collection centre for cooling as soon as possible; handle can gently
Post-milking cleanliness	Rinse all milking utensils in cold water, wash them with detergent and brush in hot water, rinse again in cold water, then rinse in disinfection or very hot water and place upside to drain
Reusing disinfectant	Do not reuse rinsing disinfectant solution for next milking
Drying of equipment	Leave utensils to drain on racks in a well-ventilated, clean, tidy place
Disease treatment	Use indicator paper or California mastitis test to detect sub-clinical mastitis; treat on same day as detected
Clinical mastitis treatment	Empty inflamed teat out every 2 hr, leave antibiotic in teat for 8 hr

14.3.1 Health and personal hygiene

People suffering from contagious disease, respiratory problems (such as a cold) or intestinal problems (such as diarrhoea) shed bacteria at greater than normal rates. To guard against the spread of disease, it must be assumed that everybody is potentially a disease carrier. People should not handle milk or hand milk cows if they have:

- sore throats or upset stomachs (diarrhoea and/or vomiting)
- skin infections (boils, septic pimples, rashes, etc.)
- heavy colds or fever
- any disease that may be transmitted by contamination of milk and dairy products.

To reduce the likelihood of contamination, anyone handling milk should be aware of bad or unconscious body habits and avoid the following:

- scratching any part of the body, face, nose, mouth, ears or hair
- coughing or sneezing directly onto milk or product
- touching, picking or squeezing pimples, boils or sores
- using spittle as a lubricant when hand milking (use hand cream if necessary)
- tasting milk or dairy products by using the fingers or a ladle that is returned to the product.

The key to preventing bacterial contamination of milk and dairy products is to keep the hands clean because they are by far the most common mode of contamination.

Hand washing. The procedure for washing hands is simple, yet is often not carried out correctly. Wash hands and forearms by:

- pre-rinsing to remove dirt and grime
- washing in a rich lather using soap and water
- brushing under the nails
- rinsing then drying with a disposable hand towel
- some Western dairy farmers routinely use disposable gloves at each milking.

Smoking. Do not smoke tobacco or any other substance when handling milk. Smoking causes:

- direct contamination of food by ash or cigarette butts
- coughing
- contamination of food from fingers touching lips while smoking
- health issues for non-smokers (the effect of passive smoking is well documented).

Protective clothing. The use of hair coverings and gloves is recommended to keep hair and skin flakes out of milk and to prevent cross contamination occurring.

14.3.2 Environmental hygiene

Environmental hygiene relates to both internal and external environments. The external environment are those areas outside of the production area, while the internal environment refers to areas where milk and dairy products are produced, packaged and stored.

Environs. Sealed, graded and drained roads and grassed areas around the dairy and cattle housing will cut down dust contamination of the dairy. Adequate drainage of the dairy and housing is essential. Effluent (manure, mud and cleaning water) must be disposed of effectively. Regular (once or twice per week) removal and spreading onto farmland or pasture areas should be undertaken.

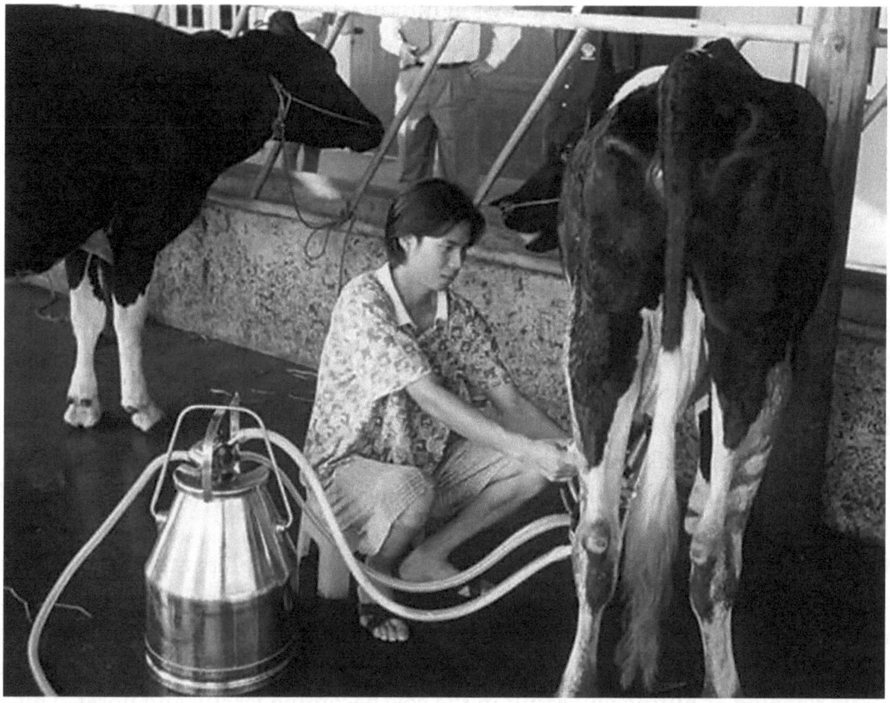

A well-managed milking system in Thailand.

Water supply. Sufficient quantities of clean drinkable water must be available for cleaning operations and a drinking supply for stock.

Rodents. Control can be a combination of the reduction of food and habitat, vermin proofing of structures, trapping and prudent use of chemical baits.

Flies, cockroaches and other insects. Insect pests carry bacteria and can transmit many diseases to humans. Control is the elimination of breeding and feeding places. Insect-deterrent lighting, electrocuters and sticky pads also have their place. Reliance on insecticides alone should be avoided.

Birds. They carry *Salmonella* and will contaminate buildings, water supplies and surrounds with faeces. Birds are attracted to dairy premises by the presence of cattle feed.

Animals and housing. These can be a major source of contamination if not kept clean. Grooming and clipping is important in reducing contamination from hair and dust on the animals. This is particularly important when hand milking. Animals should always be handled quietly and gently to avoid upsetting them. Upset animals always result in more dust and manure. Calves, young heifers or other animals (ducks, chickens, etc.) should not be housed in, or have access to, the dairy.

People. The potential for workers to bring contamination on clothing, footwear and on the person is always a possibility.

Internal environment. The dairy should be a dedicated building for milking. Cattle feedstuffs, chemicals or medications should not be stored in the dairy (except for cleansers and sanitisers). If feeding concentrates at milking, they should be stored outside the dairy and only brought in to feed cows at milking. Roughages should not be fed during milking. Dust and spillage of feedstuffs should be minimised.

The actual milking and milk-handling areas of the dairy need special attention. The floor should be an impervious surface (such as concrete), and maintained in good repair. It should be kept clean both during and at completion of milking. Dairy equipment and facilities such as wash up troughs and drainage racks need to be constructed of non-absorbent, corrosion-resistant materials, such as stainless steel.

14.3.3 Milking procedures and milk handling

Dairy/milking area. The milking area should be kept clean. Thorough cleaning following each milking session is essential. Floors should be swept and/or washed with water so as to be visibly clean.

Equipment. Teat cloths, buckets, stools, waste/dirties bucket, strip cups, teat dip and milk storage equipment all need to be clean and ready to use. Any equipment coming into contact with milk, such as buckets, milking machines and storage vessels, should be sanitised and allowed to drain for at least 15 min prior to use.

Cow handling and preparation. Good cow handling and preparation should begin before she is brought to the milking area. Gentle handling at all times is essential and, with some, grooming, if hand milking, is necessary. Slapping or striking the cow with the hand or objects such as sticks or canes should be minimised. Consistent use of even moderate slapping or hitting will result in the cow becoming fearful and upset. This will negatively effect milking because it interferes with the 'let-down' response.

Offering concentrate can be an excellent way to entice cows into the milking area, but it should be done consistently at every milking. Only give a set amount of concentrate at each milking. Offering cows more some days and less at others is likely to upset them.

Foremilk stripping. Teats should be striped into a strip cup to check for mastitis/abnormalities in the milk for at least a month (and preferably longer) into the lactation.

Teat preparation. Teat cleanliness is essential prior to milking. Dirty teats should be washed with clean running water (at low pressure) and then dried with clean, individual towels (paper or cloth). If the udders and teats are consistently dirty at each milking, the cows' environment requires closer attention.

Milking. At least 30 seconds from the start of teat handling should be allowed before hand milking. This provides adequate time for the 'let-down' response to occur. Milking should be complete within 5–7 min. Hand milking should be quick and gentle using a squeezing action rather than a pulling or stripping action. If milkers insist on using lubricants, then only allow them to use hand cream.

Hand milking technique. The most common milking technique is 'hand strip' milking for the entire milking cycle, rather than the accepted, and presumed better, alternative 'hand squeeze' method. Udder and teats are washed with a cloth but left wet to facilitate lubrication for strip milking. The cow is strip milked either one-quarter (one handed) or two-quarters (two handed) at a time. The 'hand squeeze' method more closely mimics the natural calf sucking reflex, which seals the top of teat with the lips and squeezes the teat with the tongue. The strip milking technique, combined with a non-dry udder, leads to additional bacterial loading to the milk in the bucket as the water on the teats acts as a vehicle and stripping helps push the water into the bucket.

Strip milking will also leave the teats moist, loosen the bacteria around the teat skin and provide a vehicle, via water and stripping, to bring bacteria close to the open teat orifice. The absence of teat dipping means there are always bacteria readily available to enter the teat post-milking. Strip milking is much more likely than squeeze milking to cause epidermal micro erosions that harbour mastitis bacteria.

14.3.4 Post-milking

Milk filtering. Immediately following hand milking, the milk should be filtered into a clean, sterile storage container. The filter cloth should be thoroughly cleaned in detergent and sanitiser then dried in the sun.

Milk storage/cooling. It is vital that milk is chilled to below 3–4°C as soon as possible after milking.

Milk transport. For small holders without refrigeration, milk needs to be transported carefully and as soon as possible after the completion of milking and cooled immediately at the milk collection centre. Heat, light, excessive movement and time all cause deterioration in warm milk. Transport containers should be clean, sanitised and able to be sealed with lids. They should be made of food-grade materials, which are capable of being cleaned and sanitised properly.

14.4 Milking machine function and maintenance

14.4.1 Milking machine components

Vacuum pump. This is an air pump or air compressor that removes air entering the milking machine through the various components such as the teat cups and pulsators. The pump must have sufficient capacity to remove all this air, as well as enough spare capacity (reserve or effective reserve) to maintain the vacuum at a pre-set level. It must also be able to return vacuum levels quickly to these levels following air admissions, which occurs, for instance, when cups are removed or fall off a cow.

Regulator. This simply regulates the vacuum level in the milking machine. Desirable features for a regulator are an ability to quickly regulate changes that occur in the vacuum levels of the milking machine during operation. Vacuum levels in milking machines are generally set between 40–50 kPa depending on the type of machine and equipment set up. A good rule of thumb is that vacuum levels be set as low as possible. Cup slips indicate that the level is too low.

Pulsators. These involve a valve mechanism that alternates vacuum and atmospheric pressure in the chamber between the teat cup liner and the shell of the teat cup. The cyclic pressure changes cause the liners to move within the teat cup.

The pulsation rate is the number of times the teat cup liners complete a pulsation cycle in 1 min. An ideal range for the pulsation rate would be around 60 cycles/min ± 2 cycles. Acceptable rates can go as low as 50 cycles/min but should not exceed 62. The pulsation ratio refers to the portion of time of the vacuum phases that occur during each pulsation cycle. A pulsation cycle is divided into four main phases:

- The increasing vacuum phase (**a phase**) is when the vacuum level in the chamber between the liner and the shell is increasing from atmospheric pressure to machine vacuum level. Milk begins to flow from the teat end during this phase.
- The maximum vacuum phase (**b phase**) occurs when full vacuum is achieved in the chamber. During this phase, the liner is fully open and milk will be flowing from the teat end.
- The decreasing vacuum phase (**c phase**) occurs when the pulsator valve opens the chamber to atmospheric pressure causing the liner in the teat cup to begin collapsing under the vacuum within the liner. Milk flow will cease during this phase.
- The minimum vacuum phase (**d phase**) is when the chamber is at atmospheric pressure. The liner is fully collapsed and exerting maximum pressure on the teat.

14.4.2 Testing milking machines

With all types of machines, the pump should be run for 5–10 min before testing commences. This allows the pump and machine components to reach full operating efficiency.

While the machines are warming up, a visual inspection can be carried out.

Check that:

- the guards over belts and pulleys are in good order
- the electrical wiring and installation is not worn and cut-out devices are operable
- the vacuum pump oil feed and mounting are satisfactory with little vibration during operation
- the vacuum regulator is mounted according to manufacturer's recommendations and is subject to vibration
- the vacuum gauge is fitted between the regulator and the first pulsator and were it can be read easily by the operators
- the rubberware is in good condition and free from cracks and holes
- the air admission holes on the claws are not blocked
- all air filters on regulators and pulsators are clean.

Pulsators. The important characteristics to test for and measure are the pulsation rate, ratio, and actual time of the d phase. Pulsation rate is generally acceptable in a range of 50–60 cycles/min.

Pulsation ratio characteristics that are monitored closely are the d phase and the b phase. The d phase should reach atmospheric pressure for at least 15% of the cycle and be not less than 0.15 seconds in duration. The d phase, also called the 'squeeze phase', is very important in maintaining good teat health. The b phase should not be less than 30% to ensure an optimum milk removal rate from the teat.

If there is more than one pulsator, a tolerance of 5% difference should not be exceeded in pulsation characteristics between all pulsators on the plant.

Vacuum levels. The working vacuum level is controlled by the vacuum regulator. Although there is no formal specification for the working vacuum level for most systems, a level of between 40 and 50 kPa is acceptable. The capacity of a vacuum pump to maintain vacuum levels is critical. As a guide for bucket milking plants, a basic allowance of 150 L/min, plus 60 L/min for each set of teat cup units, is a standard pump capacity. This standard allows for all the consumption requirements of the milking machine, as well as providing an allowance for an effective reserve. All effective reserve measurements at testing are recorded at a standard 50 kPa. The vacuum gauge can be checked for accuracy over a range of vacuum levels from 40 to 55 kPa and the working vacuum level set for the machine should be marked on the gauge.

14.4.3 Routine maintenance

Listed below are some basic maintenance requirements to keep a milking machine in sound operating condition.

Vacuum pump. Most pumps will have oil lubrication systems that need to be kept topped up. Manufacturer's recommendations should be followed. The drive belts should be checked for wear and cracking and pulleys should be aligned.

Pulsators. Most pulsators used on bucket milking plants are integral systems operated by vacuum and are mounted on the milking machine so as to access the vacuum. Once again, manufacturer's recommendations should be followed regarding maintenance needs. Generally this requires regular cleaning of air filters, depending on how dusty conditions are, and regular maintenance of working components within the pulsator.

Regulator. Once again, large volumes of air flow through the air filter on the regulator and it should be cleaned routinely. Manufacturer's guidelines need to be followed with regard to the maintenance requirements of the regulator.

Rubberware. From the functional aspect of a milking machine, the most important piece of equipment on the machine is the liner or inflation. The inflations are the only part of the machine that comes in contact with the cow. From the day they are fitted to the milking machine, they begin to deteriorate and lose their flexibility. A good rule of thumb is to replace the inflations after 2500 milkings, using the following simple formula:

$$\text{Replacement age (days)} = \frac{2500 \times \text{number of claws}}{\text{herd size} \times \text{milkings/day}}$$

Therefore for a 40-cow farm milking twice daily, with one mobile milking machine with two claws, the rubberware should be replaced every (2500 × 2)/(40 × 2) or every 62 days.

Apart from functional aspects of the inflation, the other concerns are the milk-quality problems caused by deteriorated rubberware. Long milk rubbers should also be replaced regularly (usually every 9–12 months).

15

Adding value to milk

This chapter briefly discusses the range of dairy products that can be made from raw milk.

The main points in this chapter

- Heat-treated liquid drinking milk is a common form of adding value to raw milk.
- Separating the milk fat from other milk solids produces butter and ghee, and also skim milk.
- Fermenting milk produces yoghurt and cheese.
- The by-products skim milk and whey can produce ice cream, cheese and whey-based drinks.
- A case study is presented of cheese making from a dairy cooperative in East Java.

Dairy farmers have the opportunity to add value to raw milk by producing a wide variety of processed products, to increase the return above the base price of raw milk. These include:

- heat-treated liquid drinking milk, such as pasteurising, or ultra high treatment, then adding flavours or incorporating it into other drinks
- separating the milk fat from other milk solids to produce butter and ghee, and also skim milk
- fermenting milk to produce yoghurt, which is also called curds in some countries
- fermenting milk to produce cheese with its major by-product, whey
- using the by-products, skim milk and whey, to produce ice cream, cheese and whey-based drinks
- adding ingredients to heated milk to produce a variety of sweets such as caramel. Further processing, such as adding acids to coagulate the milk, produces different kinds of sweets
- using cheese in cooking to produce high-value products such as cheesecakes.

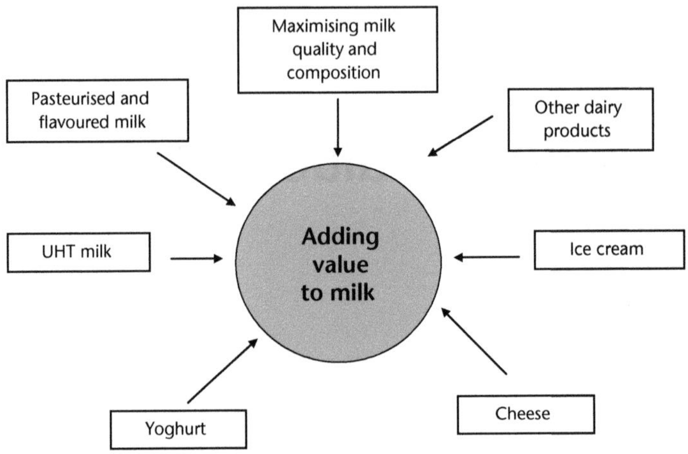

Figure 15.1. The diversity of dairy products that can add value to raw milk

Adding value to raw milk can be undertaken on an individual farm or a group of farms, using a larger, and often more reliable, supply of the raw product. Developing such cottage industries have several requirements that need to be considered. These include:

- a regular supply of quality milk that can be further economically processed
- infrastructure to support the processing at the farmer, village and regional levels
- organisational support for farmers and farming communities to develop the processing skills
- infrastructure for marketing the final products.

Pasteurising and flavouring milk in an Indonesian dairy cooperative.

The two major constraints are then the supply and quality of milk. Another factor is the existing farmer returns for the raw milk and the equivalent returns for the processed product. If milk returns vary throughout the year, depending on the supply and demand, monthly profit margins can vary. It is also likely that the products from cottage industries may not sell for the same amount as their imported equivalent.

15.1 A case study of adding value to raw milk in Indonesia

This chapter discusses value adding using an example from a project in East Java where cheese making was introduced to a dairy cooperative for sale in nearby capital cities (Moran *et al.* 2006).

It was essential to find a dairy cooperative and group of farmers who would commit to cheese making as a cottage industry, because of the considerable investment of time, training, resources and possibly short-term reduced profits.

Successful cheese making depends on using good-quality equipment and ingredients. Good milk quality is of paramount importance because the raw milk is heated, firstly, to incorporate the cheese starter and, secondly, to initiate the cheese-making process. If the raw milk already has high levels of bacterial contamination, heating will dramatically stimulate further bacterial growth leading to failure of the coagulation and cutting into curds and whey. The TPC (see Chapter 14) of raw milk destined for cheese making should not exceed 300 000 cfu/mL.

The value added to raw milk is maximised if all components of the milk can be used. That is, for cheese making, the whey has a market value either for sale fresh or as a component of another cheese such as Ricotta. Making butter or ghee could be combined with adding value to the skim milk, such as through making ice cream. Therefore, when promoting the concept of adding value to raw milk, the entire 'package' should be discussed, such as mozzarella and ricotta cheeses/whey or butter/ice cream.

Mozzarella and ricotta cheeses are ideal to introduce into cottage industries because they are heat treated during the processing, thus incorporating a sterilisation phase to ensure food safety. Mozzarella cheese is a unique Italian cheese that is mostly used as a pizza topping, hence is becoming part and parcel of affluent Asian lifestyles. It obtains its stretchy, stringy texture during a step where the curds are immersed into a bath of hot water, and worked until smooth. It is a cheese with a milk flavour and should be eaten fresh.

Ricotta, another Italian cheese, is unique in that it is made from the whey that separates from the curd during cheese making. Only fresh whey can be used when making ricotta; that is, the whey obtained from a cheese that uses rennet to set the curd. It should not be held for more than an hour before making ricotta, or it will continue to acidify and no longer be suitable. It is important that the pot used to make ricotta is deeper than it is wide: preferably twice as high as wide, to ensure a good yield.

Ricotta cheese is made from whey plus raw milk plus vinegar without additional starter or rennet. The ricotta cheese-making process is complete within 1 hr, compared with 14 days for mozzarella.

Mozzarella cheese making can be a profitable venture, in this case adding more than 50% value to the raw milk. This could increase in the future as the 'hand-made' cheese

The middleman is an integral part of informal milk marketing in Pakistan.

becomes more consistent in supply and value, thus improving its competitiveness with imported products in large cities in Asia.

Whey has long been considered a waste product and it is often discarded or used as a cattle feed. However, whey must be seen as a by-product with valuable nutrients, whose manufacture or proper utilisation has very high economic importance. It can be fed directly to cows or replace some of the milk fed to calves. Its efficient utilisation for humans requires sophisticated machinery, but fresh non-processed whey makes an ideal health drink. However, its bland taste and short shelf life have limited its acceptance and distribution.

Whey can be blended with fresh fruit or with other local products. When sweetened to taste and still warm, whey makes a delicious health drink that could be offered to women and men, as a nutritional supplement for their children, as they deliver milk to the milk collection centres associated with cheese plants. Such a procedure may require a promotion campaign, but drinking whey can form an important component of return from cheese making.

15.1.1 Constraints for cottage industries to add value to raw milk

This ongoing project created considerable interest in East Java and highlighted a variety of constraints to a successful cottage cheese industry. These include:

- the technical skills needed in cheese making to produce a consistent, high-quality product
- the supply of quality raw product
- equipping the cheese factory for operations (including buildings)
- a basic concept of hygiene on farm
- a stringent hygiene plan for the cheese factory, to include standards in clothing (including head covering), personal habits and health status
- skills to precisely monitor the processes involved in making cheese
- the need to develop taste discrimination as part of quality control
- skills in recording factory management, inputs and outputs, to allow calculations of production efficiency and to use these in troubleshooting problems
- initiating a quality control program to ensure product consistency and food safety
- developing diagnoses and action plans to tackle potential problems in quality control
- organisational support, such as sourcing government grants and loans
- provincial support in fostering the development of new cottage industries
- business management skills to source day-to-day requisites for cheese making
- preparedness to invest in an infant industry, with both resources and reduced short-term profit margins
- sufficient profit margins to justify the investment of time and resources
- a commitment by dairy cooperative management that improved milk returns will be returned to the farmers
- the scale of the cheese making operations to reduce unit transport and handling costs, and to better use the major cheese by-product, namely whey
- skills in market intelligence and promotion of a 'new' product, namely hand-made cheese
- the development of networks for marketing the finished products
- skills in preparing dishes to extend the diversification of the new products
- possibly incorporating other value-added dairy (and non-dairy) products in the product range, such as yoghurt, butter, ghee, ice cream, whey and caramel
- possible utilisation of biogas to reduce heating costs.

16

The business of small holder dairy farming

This chapter discusses the key aspects of good farm business management for the SHD farmer.

The main points in this chapter

- Production technology is just one of the skills of good dairy farm business managers. Other skills include monitoring farm production costs, obtaining any necessary credit and finances, marketing the end product, managing labour and gathering information relevant to milk production.
- There are many providers of services for SHD farmers from whom to seek information, much of which is free.
- Such service providers should allow the farmer to outsource as much of his farm inputs as possible, as long as it is viable in an economic sense. The farmer can then concentrate his efforts on what he probably does best, namely convert feed (forages, and concentrates) into milk.
- Input costs in SHD farms can be broken down into four categories: two variable (herd and shed, and feed costs) and two overhead (cash and imputed or hidden costs). Imputed costs are sometimes called 'hidden' costs, which include unpaid family labour and depreciation on farm assets.

Dairy farmers are business managers, irrespective of the size of their milking herd. A successful business is based on a good understanding of the technology underlying the production of the end product – in this case raw milk – and the ability of the manager to run the day-to-day operations at a profit and to make astute decisions regarding investments in its sustainable future. The scale of operation is generally limited by personal asset worth, but even SHD farmers can make good financial returns on their dairy enterprise.

Every day we intuitively manage our personal assets when we make household business decisions at shops, schools and in the wider community. It is the same with the

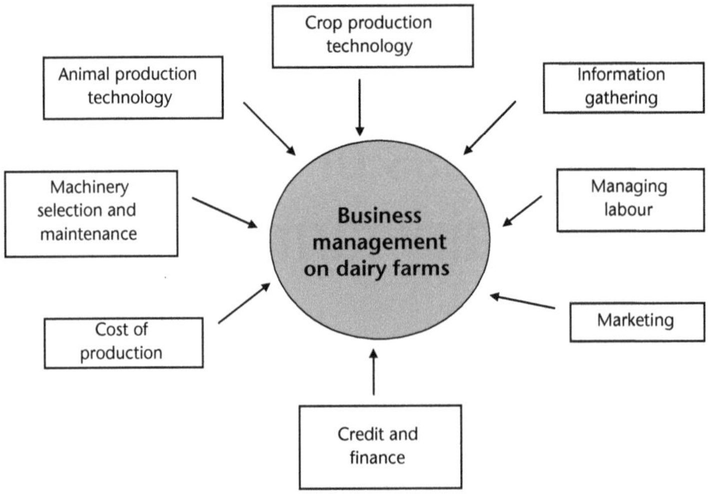

Figure 16.1. The key factors to consider in managing dairy farm businesses

SHD farmer when deciding on today's livestock feeding program, next week's crop agronomy program or the optimal herd size for next year's likely farm gate milk price. Such decisions are based on the elementary frameworks of farm business management.

Brick pillars and roofs are economic in Pakistan.

Most farmers intuitively think about farm costs and returns. However, greater use should be made of ways to make them become aware of the relative importance of all their financial inputs, in terms of their contribution to the cost of production (COP) per kilogram of milk produced on the farm. In addition, when contemplating changes in their routine farm practices, such proposed changes should be appropriately costed to allow them to make more meaningful and timely decisions.

The performance and sustainability of any dairy value chain in the tropics depends on the continued supply of raw milk from the small holder farmers. Economic pressures, such as those experienced by dairy industries throughout South and East Asia, require each farmer to be more aware of their individual COP. Without such skills, farmers cannot prioritise their management decisions to address the high cost items of their production systems.

In addition, better knowledge of farm business management allows support organisations to more clearly define the key drivers of profit on small holder farms. This information can be used to develop regional and national strategies for government departments and national dairy organisations, such as those overseeing producer-driven dairy cooperatives, to routinely evaluate and update their industry policies.

16.1 What is good farm business management?

So, what encompasses good farm business management for small holder in the tropics? Makeham and Malcolm (1986) summarised the essence of successful small holder farming in the tropics. Of the eight key areas of knowledge they listed, five use skills in business management (in italics) as follows:

- crop production and protection
- animal production
- machinery selection and maintenance
- *economic aspects of farm management (cost of production)*
- *credit and finance*
- *marketing*
- *managing labour and communications*
- *information gathering.*

From lists such as this, a set of key task areas for good dairy managers can be selected, and within each task area, progress can be quantified through developing a series of key performance indicators (KPIs). The following sections list key tasks for three major management areas on any tropical dairy farm.

16.1.1 Production technology

- Prepare the land then plant, fertilise, weed, water (in some situations) and protect the crop (likely to be a forage crop).
- Harvest, store and market the crop (through livestock rather than in the market place) to get the best return with minimum waste.
- Feed animals properly, prevent disease outbreaks and recognise disease symptoms.

- Achieve high reproductive and survival rates.
- Obtain or produce nutritionally correct feed at the optimum (generally lowest) cost.
- Provide the right housing for effective production, protection, hygiene and harvesting of the animal product.
- Where machinery is involved, choose the most appropriate types for the job, ensure they are properly maintained and serviced, and, when necessary, find a good mechanic.

16.1.2 People skills

- Have harmonious relationships with farm workers by giving them a reasonable amount of responsibility.
- Be interested in the welfare of people working with the farmer.
- Know how to establish a clear chain of command so that each person knows to whom they are responsible and does not have several bosses telling them what to do.
- Set up a system of supervision to ensure the work done is of a proper standard.
- Create a system of communication and involvement, so that all workers know what progress is being made in achieving goals and objectives of the farm operation.

16.1.3 Business management

- Use specialist advisers to help analyse the important physical and financial aspects of the farm business.
- Through appropriate records, and other relevant information, work with an adviser to produce annual farm plans, together with budgets, aimed at producing as much food and money as they need or are able to produce.
- Prepare plans of action in case of abnormal seasons and/or price.
- Plan well in advance so that all inputs are available when required, and in the correct quantities.
- Prepare physical and financial reports at regular intervals, which are timely, accurate, relevant, brief and clear for the persons who control the farm.
- Determine the most favourable forms of credit that can be obtained for different activities.
- Develop good honest working relationships with bankers, financiers or other credit managers.
- Have the ability to prepare realistic applications and finance budgets to obtain credit.
- Have the ability to know when borrowings are too great to be repaid from farm income.
- Assess the different ways of preparing and selling the farm products.
- Work out the best way(s) of marketing (assembling, preparing, transporting and selling) to return the greatest long-term benefit.
- Be able to obtain relevant information on any problem quickly. Information sources could be other successful farmers, extension agents, private agribusiness companies, research workers, libraries, teachers and friends.
- Develop effective thinking and reasoning skills, which should be combined with common sense and even mini 'trial and error' experiments.

16.2 Making the best use of service providers

A good farm business manager should know precisely what he (and his employed and family labour) has to do on his farm to generate profit, how to find out from others what can be done to improve his profit margins and when to seek outside expertise (labour as well as ideas). Small holder farmers generally depend on themselves and their family for labour, but there are numerous sources of 'good ideas' available, many of which are free. Such personnel are called service providers and they include suppliers of services and equipment, as well as good ideas. All the service providers in the following list may not be readily available in every SHD farming community, but, if there is sufficient demand, they can be found, particularly for farmers belonging to a dairy cooperative. In many regions, there are dairy equipment suppliers (resellers) or private consultants who can advise on various aspects of farm management. Service providers have a diversity of roles to improve the technical capabilities and the decision-making skills of dairy farmers both small and large scale. These include:

- **Sampling and analysing soils for essential plant nutrients.** This is a routine service used by many Western dairy farmers to plan fertiliser programs for their forage crops. It is frequently undertaken by the fertiliser agent in Asia who mainly deals with farmers growing plantation or other cash crops.
- **Selection of most suitable forage crops for the soil type and local climate.** Agronomists working for dairy cooperatives or suppliers of crop seeds can also advise on the optimum agronomic practices, such as irrigation scheduling (if available), spreading shed effluent, weed and pest control or harvesting interval for improved grasses and legumes.
- **Purchasing concentrate feeds.** Cooperatives often have the buying power to bulk purchase feeds more cheaply than individual farmers. In some cases, they may be the only source of feeds, such as imported calf milk replacers or vitamin/mineral pre-mixes. Cooperatives usually include formulated rations as part of their service to farmer members. It is important for farmers to compare the cost (and milk returns) from formulated concentrates with those for a concentrate mix prepared on farm, particularly if wet by-products (which have a short shelf life) can be sourced quite cheaply.
- **Purchasing forages.** Because of low labourers' wages in countries such as Indonesia, farmers often spend many hours each day walking around the paddy fields and along the roads to hand harvest forages for their stock. In certain areas, non-farming villagers do this to supply dairy farmers either via direct sale to each farmer or to a 'grass market', where farmers can purchase their forage requirements in the market-place. Forage markets have also been established for purchasing crop by-products, such as maize (or corn) stover, delivered by trucks from distant sweet corn growing areas. Following rice harvest, cheap rice straw frequently becomes readily available. It is important to assess its likely milk response and even undertake a cost–benefit analysis to compare it with other forage sources.
- **Nutritional management of the dairy herd.** This is an essential service to get the best out of the young and adult stock in the dairy (or the dairy beef) herd. Nutritionists can advise on availability and cost of alternative feeds, both roughages and ingredients

or home-made concentrate mixtures. They can also advise on the cheapest source of liquid nutrients for milk-fed calves, such as raw milk versus calf milk replacer. Cooperatives or commercial feed mills usually employ nutritionists, but free advice can also be sought from agents dealing with feed additives or other nutritional products. Because feed constitutes the highest cost on farm (60–70% of total costs), it is important to get 'a second opinion' before radically changing the feeding program.

- **Testing feeds for nutrient contents.** There are university, government or even private laboratories that can undertake this service, usually for a fee. It is essential that managers of dairy cooperatives or feed mills routinely test the range of feeds they use, rather than simply depend on an average nutritive value, particularly for concentrations of feed energy and protein.
- **Animal health.** Dairy cooperative or government veterinarians are frequently the major (and only) source of advice on biosecurity and on veterinary drugs and procedures for small holder farmers. Cooperatives may also employ foot trimmers to routinely treat housed stock, particularly those living on concrete floors. A lot depends on the skills and knowledge of such veterinarians and farmers to ensure the most appropriate action to take regarding sick stock, and farmers should occasionally seek 'a second opinion' from other animal health service providers (such as university staff or resellers of veterinary products). Good dairy managers need to develop a thorough animal health program to plan routine and emergency protocols for all their stock. These include vaccination, drenching and other routine veterinary practices, as well as emergencies such as calf scours, heat stress, metabolic diseases or sudden deaths.
- **Best practices for breeding management and artificial insemination.** Dairy cooperatives or government livestock officers usually supply inseminators who should be expected to advise on all things with respect to breeding, or at least be able to source the relevant information or equipment. In some areas, commercial suppliers of dairy semen are also available to advise of breeding programs.
- **Milking cow performance, such as herd recording.** These service providers are important to monitor long-term changes in cow milk yields and reproductive performance and other important measure of genetic improvement. Cooperatives frequently have computer programs to facilitate such recording and help plan realistic breeding programs, whether it be herds with 10, 20 or 50 milking cows.
- **Milking machines.** The performance of milking machines should be checked at last once each year, with routine testing of their efficiency by monitoring pulsation rates and vacuum pressure, and also assessing the condition of liners and other rubberware. Checks should be made of temperatures of milk storage equipment and of hot water to clean milk-harvesting apparatus. Advice on chemicals for washing and sanitising machines, buckets, sieves and teat-washing cloths should also be regularly updated. Suppliers of milking machines should have the necessary skills and measuring apparatus. Some milk processors also employ field officers to visit farm suppliers and advise them on improving both milking hygiene and farm milk yield, to increase their supplies of quality milk.
- **Routine maintenance of farm machinery.** All machinery requires regular attention, such as grease and oil changes for farm vehicles or replacement of worn parts. Such maintenance protocols should be provided by machinery agents.

- **Milk testing.** Most farm gate milk payments are based on measures additional to milk weight or volume, such as milk composition and milk quality. Milk composition measures the concentration of milk fat, and solids-not-fat or milk protein, while milk quality measures the degree of bacterial contamination and the inclusion of adulterating agents. The unit price for milk can have a big influence on farm profitability. Because of its economic importance, milk testing is a routine undertaken by dairy cooperatives and/or milk processors.
- **Milk transport to collection centres.** As dairy regions develop and farmer population grow, farms can become more distant to milk collection centres. As with every bulk commodity, the situation arises when it is cheaper to pay someone to collect the raw milk from the farm and transport to the central location, rather than have to use farm transport and valuable time to take it, twice each day, to the cooperative or collecting centre.
- **Monitoring farm business management.** This service can be supplied by cooperative staff or government staff and by private accountants/advisers. Lending agencies may also assist, but they generally require budgets detailing farm costs and predicted financial benefits arising from their investment in any farm development program. To improve the likely success of a loan application, small holders should seek professional advice.
- **Contractors for capital improvement programs.** Because of the high labour costs, dairy farmers in Western countries have access to a wide range of contractors with specialist skills and equipment for a diversity of tasks, ranging from forage conservation, routine young stock practices, fencing, building sheds or other farm infrastructure, through to employment agencies to find the 'right person for the job'. Apart from builders to construct cow sheds, such contractors are hard to find in Asia.
- **Cooperative service providers.** Some dairy cooperatives, such as those in Thailand, provide a range of services for their small holder farmer members. These include contract calf and heifer rearing, where the cooperative has a facility to milk rear the calves using the waste milk from the milk testing laboratory, after which the heifers are group reared, mated, then returned to the farm just prior to their first calving. Some cooperatives also grow fodder crops on communal land, after which they harvest and ensile them either in tower or pit silos. The cooperative might also bulk purchase other wet material for ensiling, such as agro-industrial by-products. Some cooperatives even have feed centres where all the ingredients for total mixed rations are blended and placed in large containers, such as old 500 kg wool sacks. These are then delivered to each farmer every few days already formulated for direct feeding to his milking or dry cows. In other countries, such as Indonesia or China, the cooperative service also extends to machine milking, either using mini milkers if all the stock are housed in the one big shed or a separate milking parlour for larger herds housed at different nearby locations.
- **Service provision as part of development projects.** Over the last two to three decades, many national and foreign governments, international aid agencies and private sector/non-government organisations have initiated dairy development programs throughout tropical Asia. Such programs provide small holder farmer support through the provision of technical services (usually at subsidised or nil cost),

credit (for specific aspects of farm management or infrastructure) and training. This is still continuing and will for many years. Astute dairy farm managers take advantage of such opportunities.

The major role of service providers is then to allow the farmer to outsource as much of his farm input as possible, as long as it is viable in an economic sense. He can then concentrate his efforts on what he probably does best, namely convert feed (forages and concentrates) into milk.

16.3 Breaking down costs on small holder dairy farms

The actual costs of producing milk of SHD farms are the sum of two components:

1. Variable (or direct) costs, which are directly related to the farm's milk output and so to the amount of variable inputs, such as fertilisers, purchased concentrate and forages, and herd costs.
2. Overhead (or fixed) costs, which are not directly related to the amount of milk sold by the farmer, because they must be paid whether or not anything is produced. These include land rent, government land taxes, loan repayments and other finance costs, and living expenses. Labour costs can be categorised either as overhead or variable, but in this manual they will considered as overhead costs.

For the purpose of calculating the cost of production (COP) of milk production on SHD farms, costs can be broken down into four categories: two variable and two overhead (Moran 2009b) as follows:

1. **Variable costs:**
   - herd and shed costs, to maintain the entire dairy herd and to harvest the milk
   - feed costs, to feed the milking herd.

   The more milk produced and the bigger the dairy herd, the greater these variable costs.
2. **Overhead costs:**
   - cash overhead costs, which involve actual payments, such as for employed labour and interest on borrowed money, rates and other farm administration costs
   - imputed overhead costs, or hidden costs because no cash changes hands. Family labour is the classic example where the farmer and his family work the farm, but all too frequently don't pay themselves for their labour. Depreciation of farm equipment is another imputed cost, which only becomes obvious when the equipment must be replaced.

The question often arises, what is the farm manager worth to the business? The answer is either what he could earn if he spent that time being paid to do other work (that is, the opportunity cost of his farm labour), or what it would cost to employ someone else to do his job. With regards the latter, because it requires more skills to manage a large dairy herd (say 100 cows) than a small one (say 10 cows), the bigger the herd and the more complex the job, the greater should be the manager's operator's allowance or imputed labour costs. This is rarely considered in Asian SHD farming.

Small holder farmers in Malaysia collecting farm data input costs.

This COP analysis includes finance costs on borrowed money. It is the choice of the farm business adviser as to whether interest on loans is included as a cost of producing milk on Asian small holder farms. Because they can constitute a major cash outlay each year for SHD farmers, they should be included in his annual financial commitments and future farm budget projections. Consequently, they have been incorporated into this particular COP analysis. This decision highlights the importance of clearly describing the particular components of any financial analyses so the reader is clear as to exactly what is and what is not included in the final 'bottom line' measure of COP. Unfortunately, this is rarely the case, leaving the reader unsure as to how such data can be interpreted and, importantly, compared with other COP estimates undertaken by other dairy farm management specialists, who may be using different methodologies.

Fuel and oil are normally included in the feed costs, although some farm economists consider repairs and maintenance of farm machinery as an overhead rather than a variable cost. It doesn't really greatly matter as long as it is only included once.

Because dairying is frequently just one of the enterprises on many small holder farms, it is important to consider only the costs relevant to and the income generated from the dairy enterprise. Such apportioning of farm finances is often not easy because labour units, machinery and farm facilities are frequently used for a diversity of farm enterprises. In addition, if feed for any dairy animals (young stock as well as adult cows) is produced from a cropping enterprise on farm, such as rice straw or maize stover, it should be given a cost to the dairy enterprise.

The total COP is then the sum of all farm costs included in Table 16.1 and Figure 16.2. Unfortunately, one still finds published estimates of COP for SHD

Table 16.1. Categorising farm costs on small holder dairy farms

Category	Costs	Detail
Variable costs	Herd and shed costs	1. Artificial insemination; inseminator, semen, drugs associated with reproductive management 2. Young stock; raw milk or calf milk replacer, concentrates and roughages and herd management to point of calving 3. Animal health; veterinarian visits, drugs, vaccines and drenches 4. Milk harvesting; rubber liners, detergents and sanitisers, maintenance of milking machines, hot water, transport to milk collection centre, cooperative commission
	Feed costs (for milking and dry cows)	1. Purchased concentrates; formulated or ingredients 2. Purchased forages; grass, roughage by-products 3. Home-grown forages; fertilisers, irrigation, processing/storage, weed and pest control 4. Machinery; fuel and oil, repairs and maintenance
Overhead costs	Cash overhead costs	1. Paid labour 2. Farm rates 3. Farm administration and insurance 4. Finance costs; interest, bank fees 5. Other; such as telephone, professional advice, office equipment, postage
	Imputed overhead costs	1. Family labour, such as operator's allowance 2. Depreciation

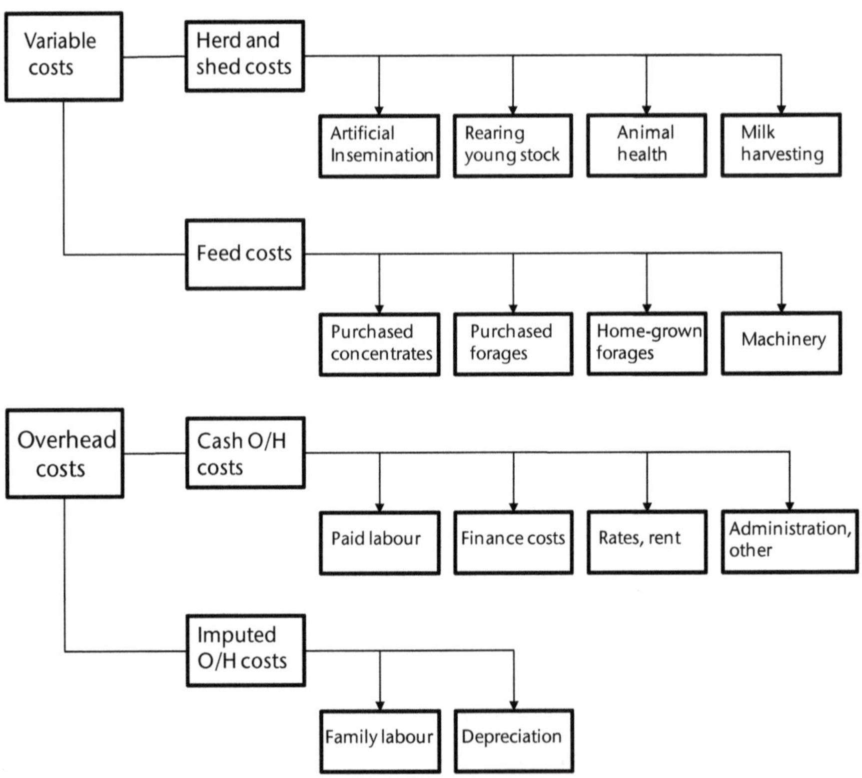

Figure 16.2. Farm costs for small holder dairy farms

operations that do not include family labour and finance costs. These create a false assessment of the true costs of dairy farming and, if used to base government policies for dairy development and even milk prices (as in some countries), they do not paint the true picture of the economics of SHD farming.

17

Conducting farmer workshops on improved herd management

This chapter presents a framework for workshops on improved herd management on small holder farms.

The main points in this chapter

- The target audiences for these workshops can be advisers or farmers, or preferably a combination of the two.
- The objectives of such a workshop should be clearly defined.
- The technical topics have been selected as the key learning outcomes from this manual.
- With farmers, the emphasis should be on visual images followed by practical examples of poor versus good farm practices.
- Farms to visit should be chosen with a specific purpose, such as to demonstrate a particular farming practice or set of practices.

This manual provides the topics for a workshop program specifically for farmers planning to import high-quality (genetic merit) stock onto SHD farms. However, farmers, being practical people, learn more from seeing and doing rather than listening and reading. This is called experiential learning. Therefore, to improve their understanding of the principles discussed in this manual, farmers need to be provided with practical examples of these improved management practices. In Australia, farmers hold regular group meetings, where they visit one another's farms, particularly those that have proven to be successful in improving adoption of on-farm practices. There needs to be a clear focus and set of objectives for these 'farmer discussion groups'.

The first phase in developing a focus could be via a workshop on improved herd management on small holder farms. This chapter discusses the planning and implementation of workshops to encourage better adoption of farm practices to better exploit the productive potential of imported dairy stock. Such workshops should include

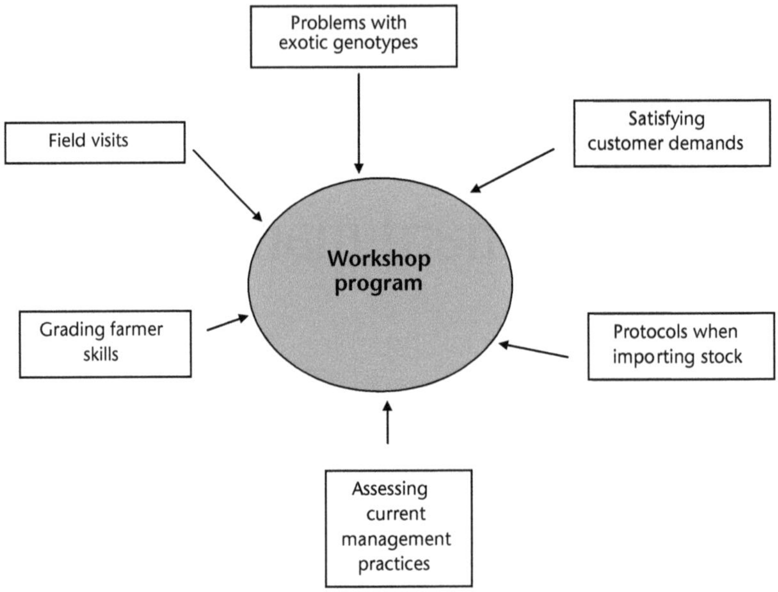

Figure 17.1. Key elements of farmer workshop program on improved dairy herd management

both local advisers and dairy farmers, so advisers can more easily develop follow up workshops and farmer discussion groups.

17.1 Planning farmer workshops

17.1.1 Workshop objectives

1. To develop more realistic expectations of the performance of imported dairy stock under traditional tropical dairy farming systems.
2. To better understand the differences in management requirements of indigenous, low production versus imported, potentially high-producing dairy stock.
3. To develop farming systems to minimise the adverse effects of the tropical environment on high genetic merit temperate dairy cows.
4. To develop a post-arrival farm management protocol for such stock.

17.1.2 Developing a workshop program

Prior to visiting local farms, the participants should be exposed to some of the theories behind the reasons for failure of previous attempts to improve the productivity of small holder farm though importing high-quality dairy stock. The following topics, discussed in previous chapters of this manual, could be considered in such classroom sessions:

Problems with exotic genotypes (Chapters 9 and 11):

- genotype by environmental interactions
- specially bred tropical dairy genotypes
- problems of confinement versus grazing.

Importing cows and heifers from other countries (Chapter 1):

- major shortfalls in management
- genetic merit of imported stock
- importing young heifers
- the renewed interest in embryo transfer technology.

Satisfying customer demand (Chapter 2):

- what the customer wants
- what the customer needs
- what the customer gets.

Developing management protocols (Chapters 1, 3, 4 5, 6, 17 and 18):

- management prior to departure (quarantine regulations)
- management during transport
- management immediately post-arrival
- transport to the new home farm
- realistic expectations of imported cow performance
- preparing the farm for the new arrival (shed, drinking water, feed, animal health)
- key aspects of herd management
- traditional survival versus improved production management
- preparing a troubleshooting guide or checklists for post-arrival protocols
- farmer and adviser training programs
- managing the progeny of imported stock.

Assessing current farm management practices (Chapter 18). This session is to prepare participants for field visits to selected farms and covers:

- feed production
- feeding management
- herd management
- housing
- milking management
- general farm management.

This would be followed by an open session on dairy production technology. The topics for this session can be based on those listed in specific expectation forms from workshop participants. It is important that these be considered because farmers (and advisers) may have particular practical issues that they hope can be solved during the workshop.

17.1.3 Practical issues

For this workshop, a series of visual images of poor versus good management should be developed (such as those in Chapter 19 of this manual). This should form the basis of any troubleshooting guide covering both post-arrival management and good farming practices. DVDs would provide better visual images than still photographs. Appendix 4 presents a glossy four-page 'easy-to-read' brochure, which could be translated into the national language of the targeted country.

Dairy advisers from Thailand collating information from a farm visit.

The workshop material should differ depending on the key audience and following discussions with extension specialists in the various countries for which this program is to be designed. As well as a simple set of PowerPoint presentations, technical workshops for government advisers and trained herd managers could include distribution and discussion of this manual. However, technical packages for all farmers receiving the imported stock should be based on simpler material such as visual images and glossy reference cards. The importance of language for communication, such as subtitles and dubbed sound track on any specific DVDs for each target country, must be taken into consideration.

Examples of expectation forms to be filled in by each participant at the beginning, and evaluation forms at the completion, of workshops have been presented in Appendix 3. These are very important, firstly because workshop participants may not all be aware of the workshop's emphasis on improved management practices and, secondly, this can help plan an 'open session' on the last day to discuss specific issues on dairy production technology.

17.2 Associated farm visits

A key element of any farmer workshop is a series of farm visits, preferably to farms with different levels of herd and feeding management, so participants can see for themselves what constitutes poor and good farm practices. It is one thing to develop a series of best management practices (BMP), but another to see attempts to adopt them and, of more importance, to understand the reasons why farmers fail to achieve them. This failure could be due to:

Advisers from Indonesia interviewing a farmer about his herd management.

- ignorance; farmers don't know about them
- lack of resources; farmers don't have the money or facilities to adopt them
- lack of incentives; farmers do not see the need to want to adopt them
- lack of service provided by government advisers or local agribusiness providers.

Unfortunately, many of these BMPs are like insurance policies in that farmers take the attitude that 'if it isn't broken, you don't need to fix it', particularly if it costs money to modify facilities or purchase additional equipment. Such issues should be discussed back in the classroom following the farm visits.

Selection of suitable farms is paramount when planning the visits. Close proximity to the workshop venue is important. Each farm should be chosen with a specific purpose, such as to demonstrate a particular farming practice or set of practices. Ideally, farmers should have good records so they can provide useful background on these observed practices, such as their herd reproductive performance and their cost of production. Clearly, selected farmers should not be intimidated by large groups of inquisitive visitors. Inviting farmers back to the workshop is useful so they can further explain their management decisions during the debriefing session.

Forming small groups of workshop participants with specific tasks allows a lot more information to be collected on farm than if the visit was less structured. These groups could meet back in the classroom to prepare a short presentation summarising their observations and interpretations.

18

Assessing current farm management practices

This chapter presents a framework for assessing current farm management practices.

The main points in this chapter

- There are a series of key performance indicators that provide realistic targets of farm performance following improvements in feeding, herd and farm management.
- The average milk production of all the lactating cows can provide a useful guide as to the adequacy of the current dairy farm management practices.
- This chapter presents six key task areas with which to assess current farming skills: feed production, feeding management, herd management, housing, milking management and general farm management.
- These could provide a framework for grading the suitability of farmers to receive imported high-quality dairy stock.

Imported dairy stock require a high level of herd and farm management to ensure they continue to grow well and remain healthy to become productive milking cows. Poor management practices can lead to low growth rates, delayed breeding, stock diseases and even deaths among imported stock, both before and after first calving.

It is important to assess the quality of management on each dairy farm prior to the introduction of these stock to help decide on the farm's suitability to be included in such a dairy improvement and importation scheme. A series of checklists to consider when assessing current herd and farm management practices is provided below. This is followed by one approach to assist in this evaluation process. Prior to working through the checklists, a new approach has been documented to provide a guide to the adequacy of current farm management practices.

The dynamic nature of dairy farming makes it difficult to develop a simple set of criteria with which to assess current management skills. The term key performance indicators (KPI) has been derived to describe a series of measures of dairy farm performance with which to provide realistic targets following improvements in feeding,

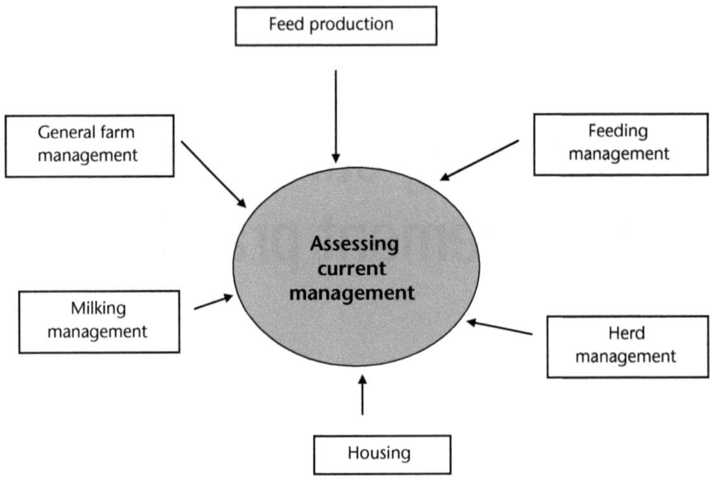

Figure 18.1. The framework to assess current farm management practices

herd and farm management. Such a set of KPIs for SHD farming were developed for one of my previous books (Moran 2009a) and has more recently been presented in a specific publication (Moran 2009b), which is freely available from the author (contact jbm95@hotmail.com). This publication contained a summary table of the 10 key measures of SHD farm performance, which is presented in Table 18.1. It is beyond the scope of this chapter to provide more details than is in this table.

18.1 Using current cow milk yields to assess adequacy of the current farm management

All these KPIs can be quantified to provide guidelines as to which ones require priority in any dairy farm improvement program. Although some are relatively easy to quantify, others are quite difficult. Probably the most simple, and most used, single measure of SHD farm performance is the average milk yield of the milking cows. The correct term for this figure is 'rolling herd average' because it is the average milk yield of all the lactating cows, which will be at various stages in their lactation cycle.

This single value provides a summation of all the important aspects of SHD farm management, so any interpretation must take into account a diversity of feeding, herd and farm factors. Accordingly, many dairy specialists may query its usefulness as a single measure of dairy farm performance. However, it is routinely used by farmers to describe their farm's performance in relation to their neighbour's farm and also in relation to production targets provided by many government advisers. In addition, it is often quoted by government officials when summarising the stage of development of their national dairy industries. Table 18.2 attempts to describe the adequacy of a dairy farm's management practices using the rolling herd average; other factors to consider are listed in section 18.1.1.

Table 18.1. Ten key measures of dairy farm performance

Measure	Description
Feeding management	
1. Stocking capacity	Is the farm carrying too many stock for the available forage supplies?
2. On-farm forage production	How much of the farm's annual forage requirements must be purchased?
3. Forage quality	Is the forage being harvested or purchased at its optimal quality for milking cows?
4. Concentrate feeding program	What is the quality of the concentrates being fed and how much is allocated per milking cow?
5. Total feed costs	Are the forages and concentrates costing too much per MJ of energy or kg of protein?
6. Milk income less feed costs	How does this compare with those of other farmers with good feeding management?
Herd management	
7. Proportion of productive cows	What is the percentage of adult cows actually milking? What is the proportion of milking cows in the entire dairy herd, expressed as a percentage?
8. Pattern of milk production	What is the peak milk yield of the herd and what is its lactation persistency (rate of decline from peak milk yield)?
9. Reproductive performance	How many days after calving do cows cycle? What are the herd's submission and conception rates? What are the herd's 100 day-in-calf and 200 day-not-in-calf rates?
10. Heifer management	What is the calf mortality rate and the wastage rate from birth to second lactation? What is the age and live weight at first calving of the replacement heifers?

Table 18.2. Interpreting the adequacy of dairy farm management from cow milk yields. Range in average herd milk yields on tropical South-East Asian dairy farms.

Herd milk yield (kg/cow/day)	Adequacy of dairy farm management practices
5	Very poor feeding and herd management and low genetic merit cows (or milking buffalo)
7	
9	Typical of many SE Asian small holder farms, even with high grade Friesians
11	Gradual response with grade and crossbred Friesian cows to improved feeding, herd, young stock and shed management. **Milk yields of 15 kg/day are considered acceptable by many government dairy advisers.**
13	
15	
17	
19	
20	Potential level in **lowland humid tropics** following improved management of body condition throughout lactation
25	High genetic merit cows in **tropical highlands** or **lowland dry tropics** with good farm management
30	Typical peak milk yields in herds with 25 kg/cow/day rolling herd averages
35	Unrealistic in SE Asia, except where all major constraints to milk production have been overcome

18.1.1 Other factors to consider

- It is important to differentiate between rolling herd averages and peak milk yields.
- Milk composition should be considered as an indicator of feeding management, for example:
  - low milk fat can indicate possible sub-clinical rumen acidosis
  - high milk protein can indicate good dietary energy intake
  - milk lactose levels are fairly constant.
- Excessive body condition is indicative of low protein diets, due to:
  - the inability of cows to partition nutrients from body reserves to milk synthesis
  - poor fertility because cows cannot easily cycle, hence conceive.
- Very poor body condition is indicative of low energy intake because:
  - high genetic merit cows preferentially partition body reserves to milk synthesis
  - cows will not cycle due to excessive weight loss.
- Herd dynamics can also indicate adequacy of dairy farm management:
  - excessive number of dry non-pregnant cows can indicate very poor farm management
  - a low percentage of lactating adult cows can also indicate poor farm management.

18.2 A checklist to assess current farm management practices

For easier interpretation of overall farmer skills, the nine farm activity areas listed in Chapter 1 have been reduced to six by:

- combining young stock, disease and reproduction management and stock welfare into one 'herd management' category
- introducing an additional 'general farm management' category to take into account the farmer's business management skills.

These skills/farm activities are described in sections 18.2.1 to 18.2.6.

18.2.1 Feed production

- size of forage production area (in m^2)
- types of forages grown on the farm (grasses, legumes, cash crop by-products)
- quality of forage at harvest
- fertiliser management (manure, inorganic fertilisers)
- forage harvest interval (during wet and dry seasons)
- fodder conservation practices (silage, hay)
- year-round supplies of fodder (dry season sources)
- number and type of stock on the farm (milking cows, dry cows, heifers, calves)
- stocking capacity (too high, optimum or too low?)
- proportion of forages from home-grown supplies
- proportion of forages sourced off farm
- types of forages sourced off farm.

Workshop participants assessing the farmer's facilities for milk rearing his dairy heifer replacements.

18.2.2 Feeding management

- typical daily allocation of fresh forages (? kg fresh forage/milking cow/day) in wet and dry seasons
- forage preparation (wilting, chopping)
- hand or machine chopping of forages
- type of concentrates fed (formulated, by-products, knowledge of energy and protein supplements)
- vitamin and mineral additives (type, others fed)
- allocation of concentrates (? kg/kg milk produced)
- measures of cow milk production (peak milk yield, current milk yield, days in milk)
- pattern of milk production (rate of decline from peak milk yield)
- body condition at calving, mating and drying off
- knowledge of cow live weights and changes in body condition scores (mating and calving live weights).

18.2.3 Herd management

- milk-fed calf management (colostrum feeding, milk feeding program, weaning age, calf mortality)

- source of milk (whole milk, calf milk replacer)
- treatment for scours (use of antibiotics versus electrolytes)
- calf pens (type, cleanliness)
- weaned heifer management (concentrate feeding, mating age, age at first calving)
- visual assessment of condition of calves and heifers
- disease management (vaccination program, veterinarian support and animal health program, drug storage, protocol for treating scours, milk fever and other diseases)
- reproductive management (heat detection, routine pregnancy testing, calving interval, conception rate)
- breeding program (criteria for selection of sires)
- lameness (locomotion test)
- mastitis (incidence, routine California mastitis tests, management and control programs)
- climatic environment (monitor respiration rates, outdoor access at night)
- structure of dairy herd (proportion of productive cows in milking herd, proportion of productive cows in entire herd)
- record keeping (board in shed, note book, computer)
- what records are maintained
- target setting (daily milk yield, long-term herd size)
- welfare issues (stock transport, any obvious problems)
- surplus stock (disposal, growing out male stock)
- other sources of dairy income (manure, excess fodder, biogas).

18.2.4 Housing

- shed design (floor slope and comfort, roof height, ventilation)
- access to water (adequate, continual)
- climate control (temperature and humidity inside shed, sprinklers, fans, varying feeding management during year)
- shed cleanliness (layout for cleaning, frequency of cleaning)
- cow stalls (tie stalls or loose housing, free stalls, sufficient lounging area)
- milk feeding calf pens (crates, deep litter, can easily separate calves, cleanliness)
- grouping of milking cows based on stage of lactation (yard for heat detection)
- pen layout inside shed (for young stock, hospital pen for isolation)
- location of milk-fed calf pens away from adult cows
- feed storage facilities (separate to stock area, bird, insect and vermin proof, mixer for concentrates, chopper for forages)
- services (insulation, electricity, water for washing down, hot water for calf feeding equipment)
- catering for personnel (suitable space and furniture for relaxing, lockable storage for drugs)
- effluent disposal (size of manure pit, how frequently emptied).

18.2.5 Milking management

- separate milking area (good layout for machine milking)

- hand milking (personal hygiene, milking method, type of and cleanliness of milking bucket)
- machine milking (cleanliness, quality and replacement of rubber liners)
- milk storage (milk cooler, stored in cool location)
- milking hygiene (hot water, soap, sanitisers, store buckets and milk cans upside down)
- milk quality and unit return (percentages of fat, SNF or protein, bacterial contamination, milk grade, expressed in local currency units/kg milk).

18.2.6 General farm management

- record keeping (board in shed, note book, computer) and sufficient office space
- extent of record keeping (cow production only, performance of all stock, all farm inputs, labour)
- systematic record keeping (accessibility, kept in logical format, is it referred to in future?)
- extent of financial records (occasionally revisited, relationship with creditors)
- knowledge of costs of milk production (? local currency units/kg milk produced on farm)
- short-, medium- and long-term plans for farm
- making good use of service providers (free information, awareness of all services)
- efficiency of using employed and family labour
- communication and management of paid labour
- milk marketing (member of dairy cooperative, considered value adding, sought all possible market outlets)
- Subjective assessment of overall farm management skills.

18.3 Grading farmer skills

Using this structured approach to assess farm management could provide a framework for grading the suitability of individual farmers to receive imported high-quality dairy stock. For example, the following set of criteria was developed for a herd improvement program in Central Java in 2008, using a grade of good, average or poor, with some objective and others subjective.

18.3.1 Objective criteria

- **Average milk yield per cow:** above 12 kg/day, good; 10–12 kg/day, average; below 10 kg/day, poor.
- **Price for existing milk produced** (because this entails milk quality and composition): above 3000 Rp/kg, good; 2500–3000 Rp/kg, average; below 2500 Rp/kg, poor.
- **Fresh forage supplies (grass, cassava leaf, etc.) in the wet season:** above 30 kg/day, good; 20–30 kg/day, average; below 20 kg/day, poor.
- **Record keeping:** daily production, artificial insemination (when), when veterinarians visit and for what disease, when and what vaccination, from where the stock originate, their predicted calving dates. If the farmer writes the details down and remembers, good; if he remembers but does not write them down, average; if he does not know or remember, poor.

18.3.2 Subjective criteria

- **Condition of cow shed:** size, ventilation, condition of floor, hygiene (where is the manure pit?).
- **Herd management:** condition of dairy cow (health, weight, body condition), condition of young stock (health, weight for age, body condition), reproduction (calving interval).
- **Milking management:** equipment, hygiene, post-harvest milk handling.

The above example was developed specifically for a dairy development program for one region in Java, predominantly based on introducing high genetic merit stock both from within Indonesia and imported from Australia (Moran 2008). Because current levels of farm management were suboptimal, the standard of the grading scores were not high. This would probably need to be revised when developing a grading score for dairy development schemes in other regions outside Central Java.

This scheme was developed for a dairy adviser workshop in which participants visited several farms to assess their current levels of farm management. This provided a focus for group discussions on the level of farm performance required for them to make the best use of imported stock and the magnitude of current constraints for farmers to achieve acceptable levels of herd performance and farm profits.

19

Tips and traps in managing high grade dairy stock

This chapter presents a series of photographs depicting good and poor management practices on tropical SHD farms

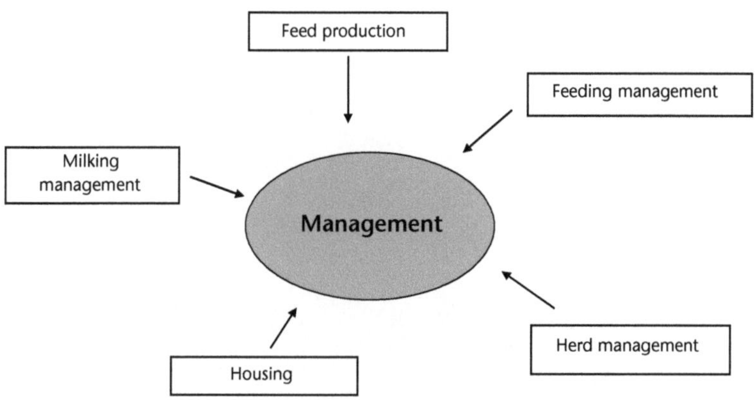

Figure 19.1. The key factors to consider when planning the milking herd's management

This collection of photographs highlight many of the good and poorly managed practices found on SHD farms throughout the world, which will influence the success of schemes to improve farm performance and profits through importing high genetic merit dairy stock.

19.1 Feed production

To optimise forage quality, Napier grass should be harvested at least every 35 to 45 days (Vietnam).

This stand of 100-day-old Napier grass will not be good quality forage (Indonesia).

Guinea grass and Napier grass require different harvesting intervals (Vietnam).

A combination of Leucaena (tree legume) and Napier grass provides an excellent forage mix for milking cows (Indonesia).

Maize (or corn) stover can provide a forage of moderate quality (Indonesia).

Alkali-treated rice straw is not a good milk forage (Indonesia).

Chopped oil palm fronds are a very poor-quality forage (Malaysia).

Banana leaves are a poor-quality forage source (Indonesia).

Napier grass requires inorganic fertilisers in addition to cow manure (Vietnam).

A by-product of monosodium glutamate production can provide a relatively cheap source of fertiliser nutrients (Vietnam).

In some places, small holder farmers can purchase forages from a grass auction (Indonesia).

Roadside grass being delivered to small holder farmers in Indonesia, but it has variable quality.

19.2 Feeding management

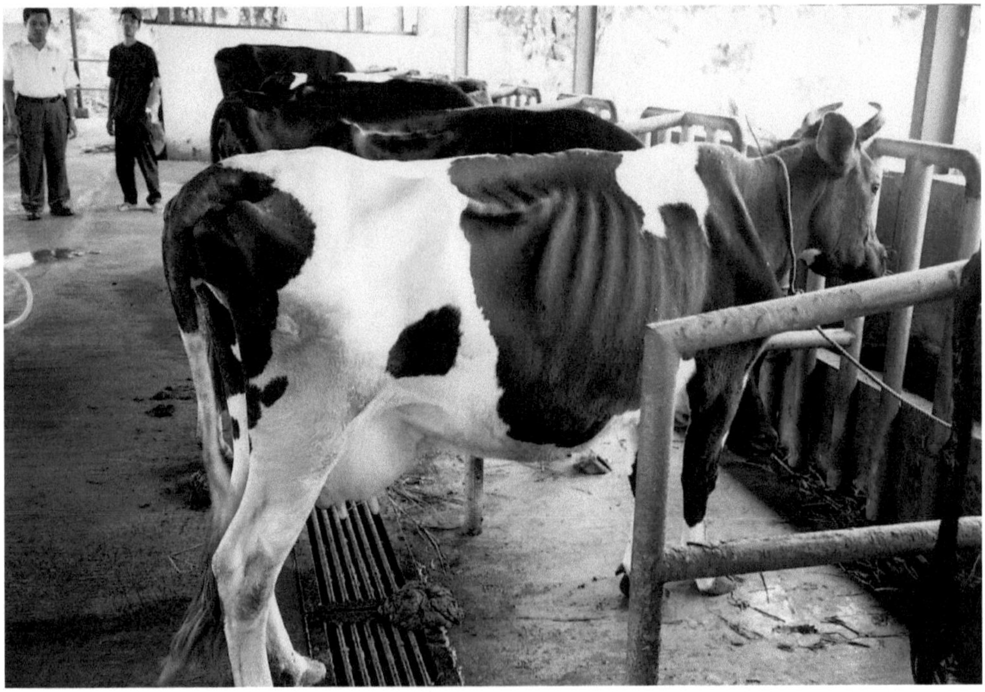

Without sufficient feed nutrients during early lactation, high grade Friesian cows can lose body condition and not cycle as well as reduce their milk production (Indonesia).

Well-managed cows in early lactation (Malaysia).

Cows also require a balanced ration throughout lactation, otherwise induced protein deficiencies can lead to overfat and infertile cows (Malaysia).

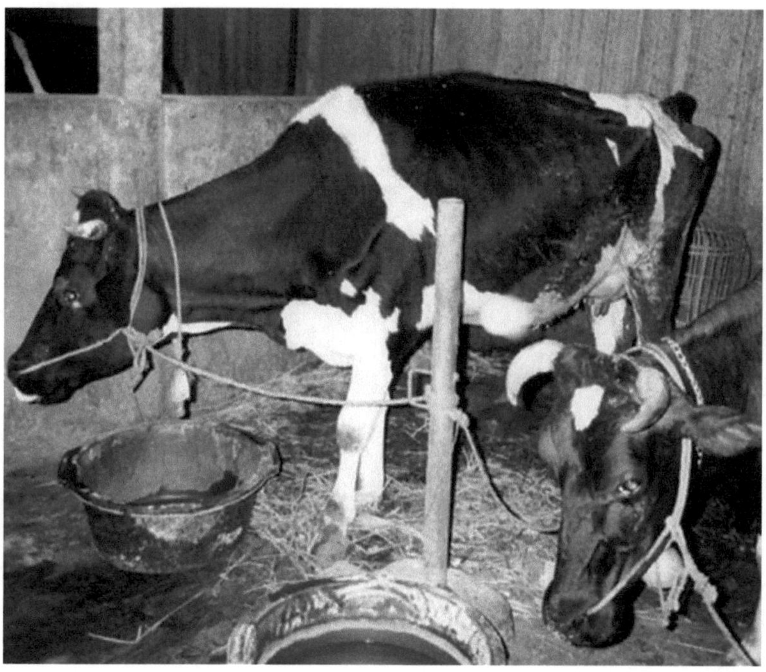

Poorly fed cows are not productive cows (Indonesia).

Supplements in addition to forages are required by milking cows, such as citrus pulp (Malaysia).

Rice straw is a poor forage for milking cows (Indonesia).

Immature Napier grass is an ideal forage source for milking cows (Malaysia).

This cow is being fed a concentrate slurry containing chopped sago palm trunks (Indonesia).

19.3 Herd management

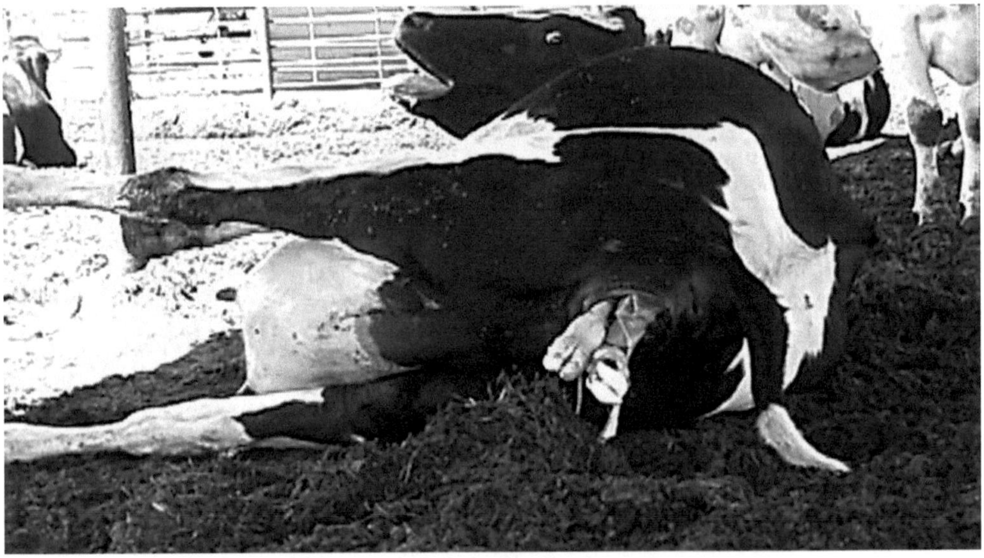

Good young stock management starts at birth by ensuring the calf is born onto a clean area.

Ideally, milk-fed calves should be reared in separate crates to minimise calf contact and allow for individual attention (Vietnam).

Deaths of young stock should not exceed 2–4%, while the incidence of calf sickness, such as with this calf, should not exceed 10%.

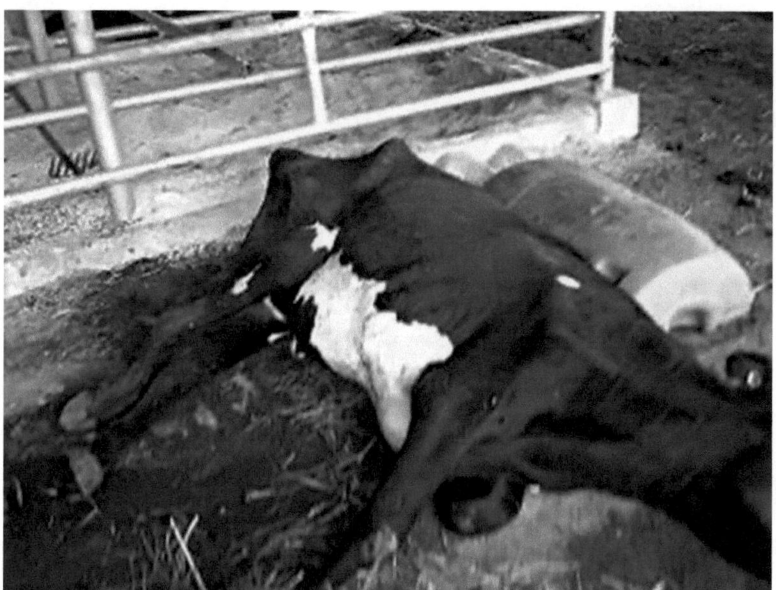

It is unlikely that this heifer will ever make a productive milking cow and should be euthanased (Malaysia).

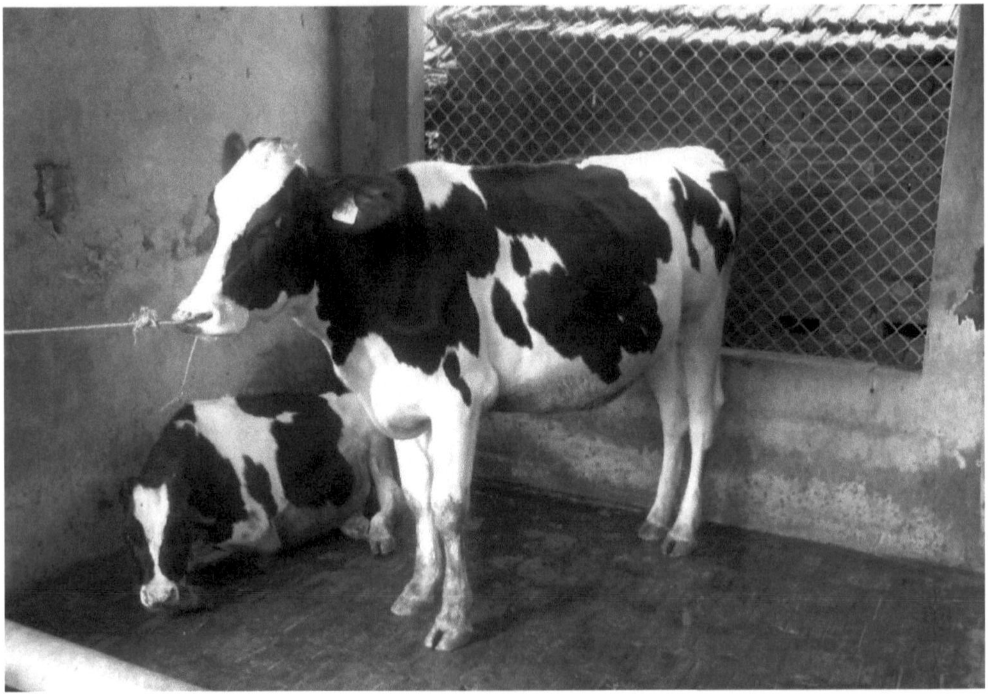

A 24-month-old heifer, 4 months pregnant. The objective of any young stock management program (Vietnam).

Heifers can appear well managed until an inspection of their teeth indicates they are 3 yr of age (Indonesia).

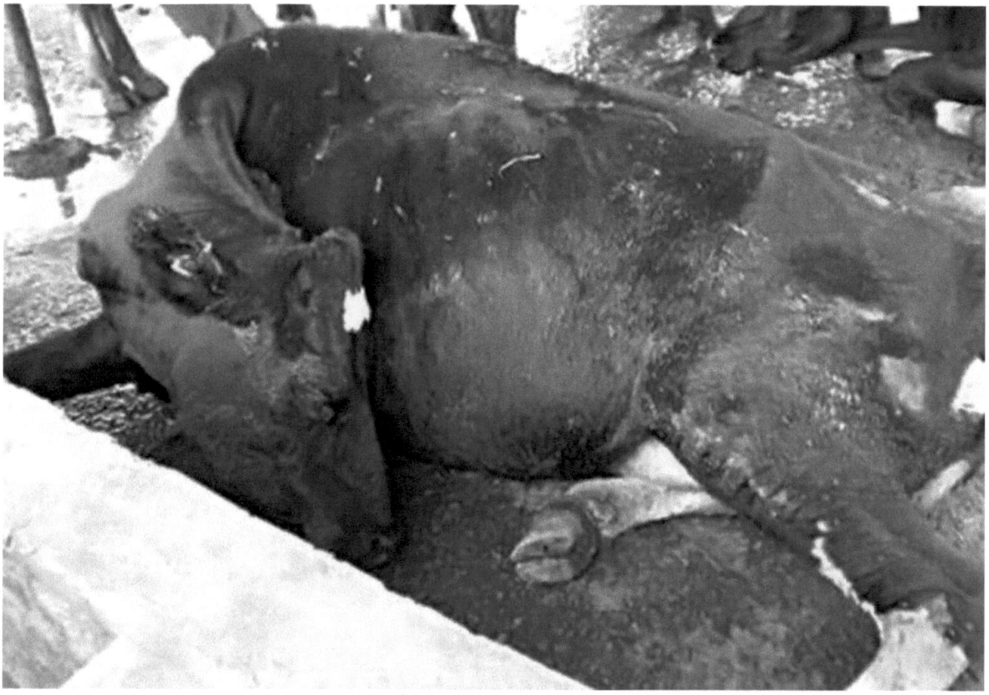

This animal requires urgent veterinary attention (Malaysia).

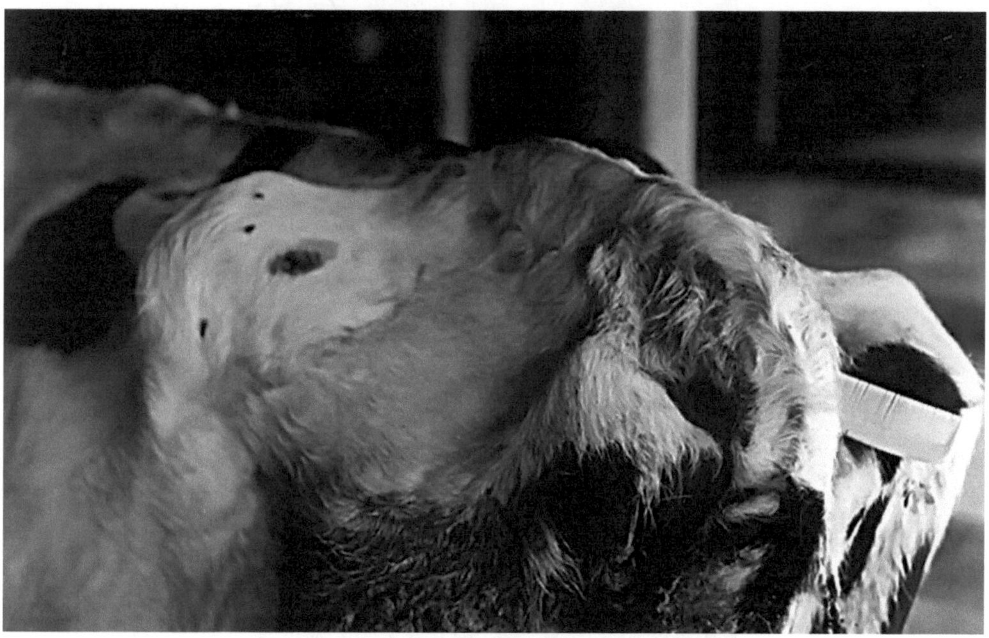

For animal welfare reasons, this animal should be euthanased (Indonesia).

This calf has been dead for at least 24 hr, clearly indicating a breakdown in farm management (Malaysia).

Sahiwals are a tropically adapted and productive dairy breed (Pakistan).

A local Indian dairy cow that is not very productive (Malaysia).

One problem in some countries is that women undertake much of the farm work but are excluded from farmer meetings (Pakistan).

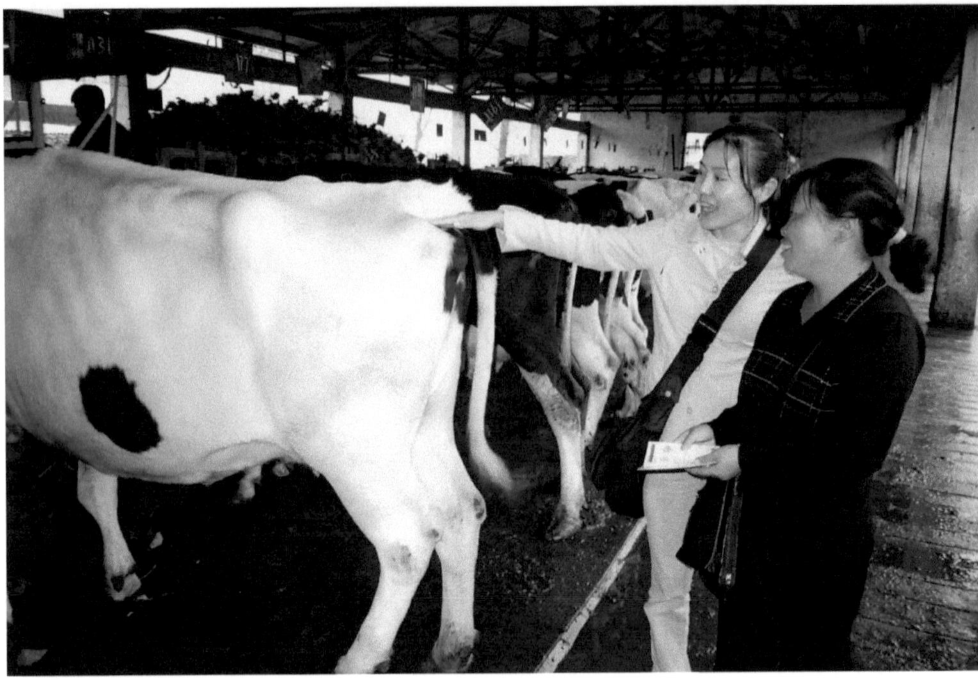

Monitoring changes in body condition is a useful tool for planning feeding programs (China).

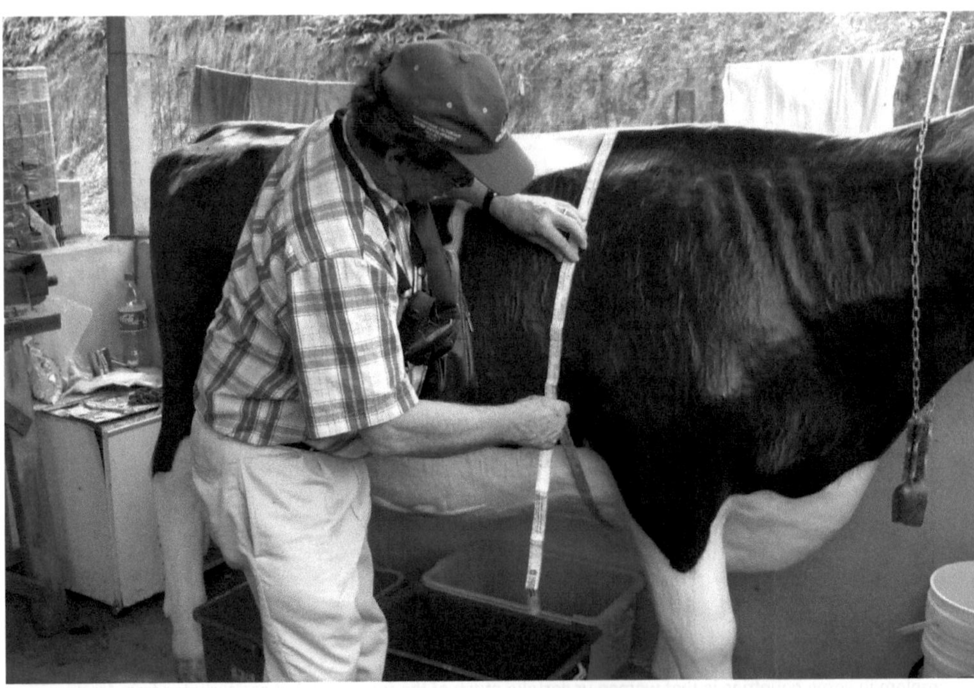

Measuring cow live weights from their chest girths is another useful tool in planning feeding programs (Malaysia).

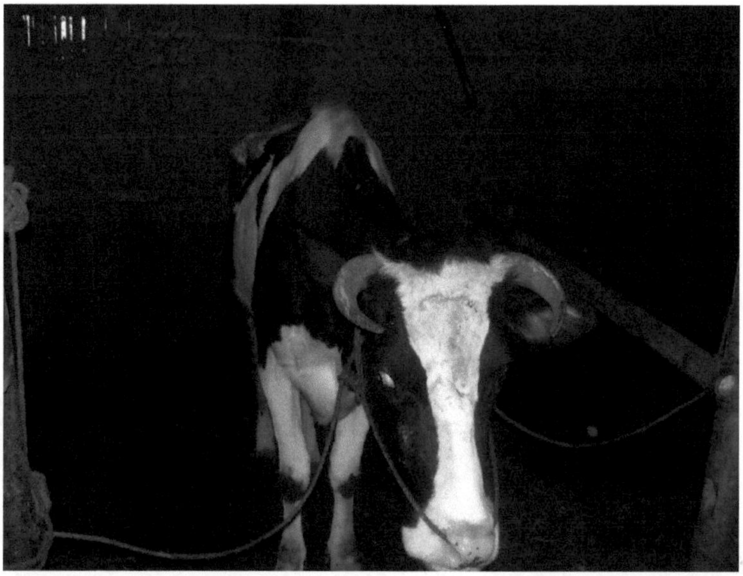

This cow is tied up all day in a darkened shed and fed a poorly balanced ration with limited forages and expected to produce more than 15 kg/day of milk (Indonesia).

19.4 Housing

Lack of shade and a forage diet based on wheat straw will severely limit the performance of these milking buffalo (Pakistan).

A good shade structure for these milking buffaloes (Pakistan).

These milking buffalo live in a dark shed for some reason not fully explained to me (Pakistan).

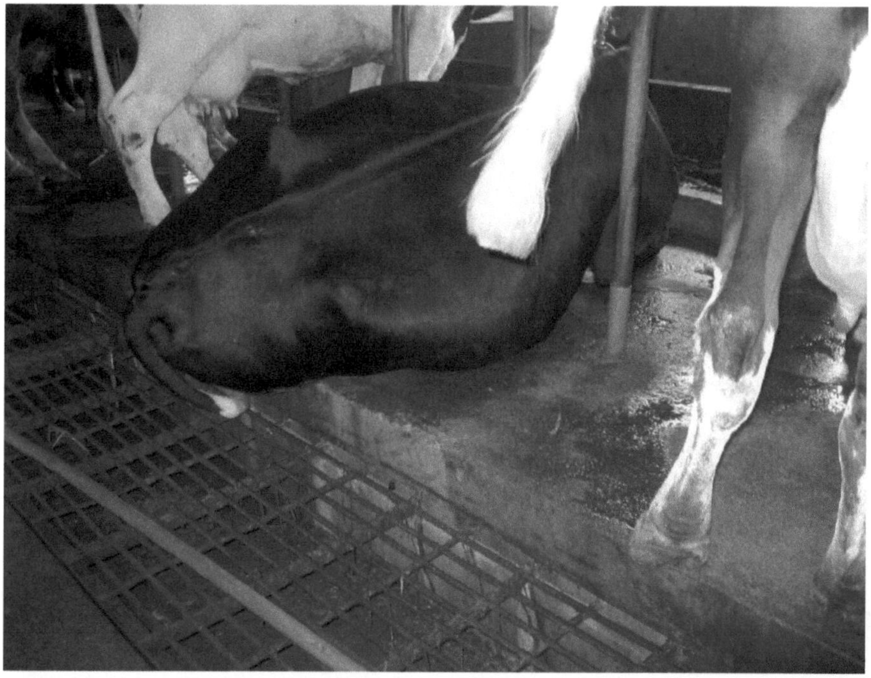

This cow has insufficient space to comfortably lie down (Sri Lanka).

Counting the respiration rates of milking cows can indicate the level of heat stress (Malaysia).

Hosing down a cow can reduce heat stress (Malaysia).

Despite the use of fans, cow performance in this overcrowded shed is severely restricted by heat stress (Malaysia).

Sprinklers in combination with fans can alleviate heat stress in milking cows (Pakistan).

Hygiene can be greatly improved in this shed (Malaysia).

The roof in this shed is too low to allow adequate ventilation (Malaysia).

Hygiene can be greatly improved in this shed (Indonesia).

This shed is a converted piggery and clearly the roof is too low (Malaysia).

For some reason, this curtain was installed to reduce cold stress in coastal Pakistan.

This is the shed for which the curtain was installed (Pakistan).

A well-designed small holder shed in Indonesia.

This shed has too many heifers in it with some animals showing traumatic injuries from overcrowding (Malaysia).

A well-constructed cow colony shed in which many farmers house their small herds (Indonesia).

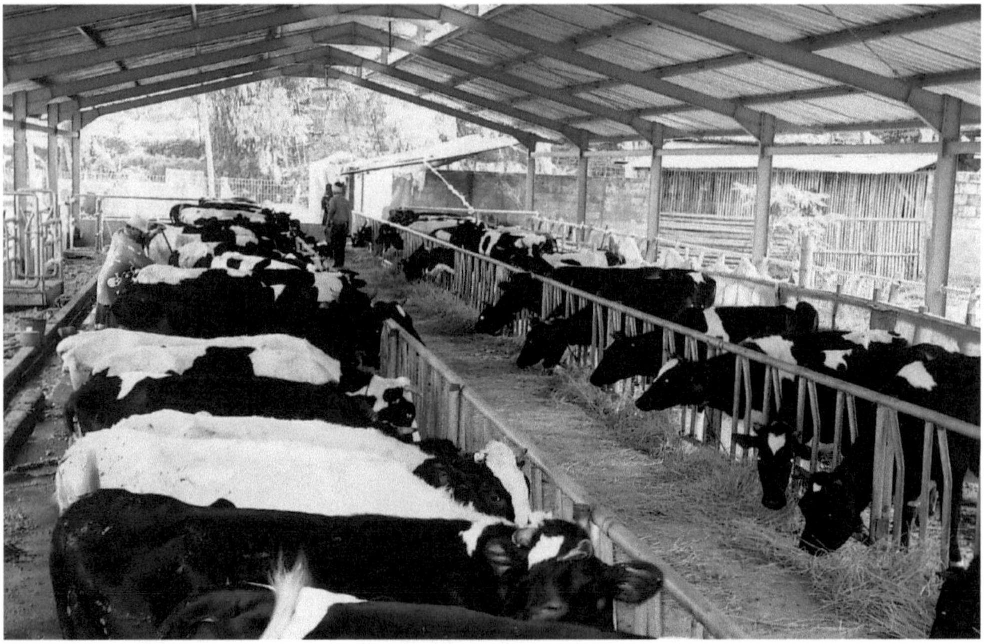

Inside the cow colony shed (Indonesia).

An individual farmer feeding his cows in the cow colony shed (Indonesia).

Shed management should be improved before this farmer is sold imported dairy heifers (Indonesia).

A well-constructed shed in Vietnam providing the manager a good overview of the stock from his office.

The stalls could have been made longer for these cows (Indonesia).

A well-designed shed under construction in Pakistan.

A well-planned free stall shed in Malaysia.

An attempt to improve ventilation, using a table fan, in a low-roofed shed in Vietnam.

An all too common problem of slippery floors due to poor hygiene in a shed in Malaysia.

19.5 Milking management

Poor milking hygiene will reduce unit returns for this small holder farmer (Indonesia).

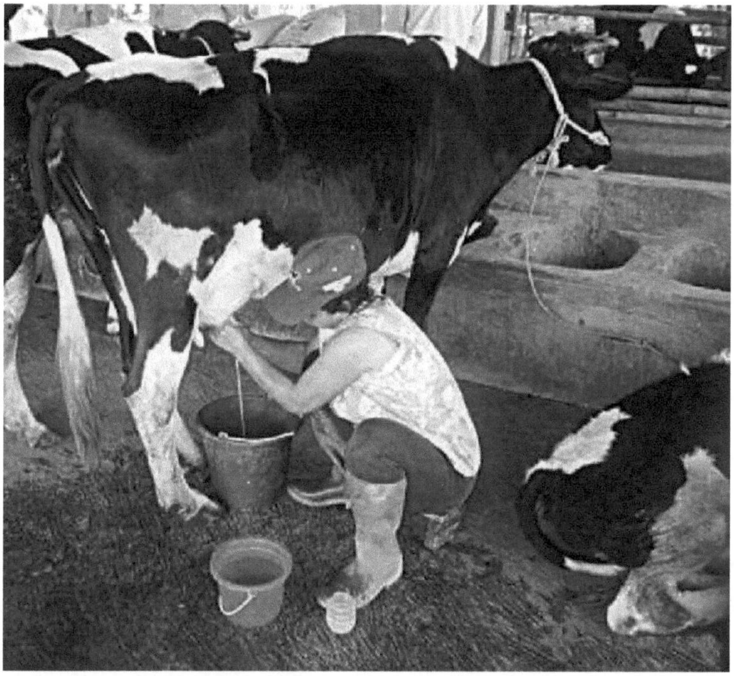

The milk bucket used by this farmer will be difficult to sterilise (Indonesia).

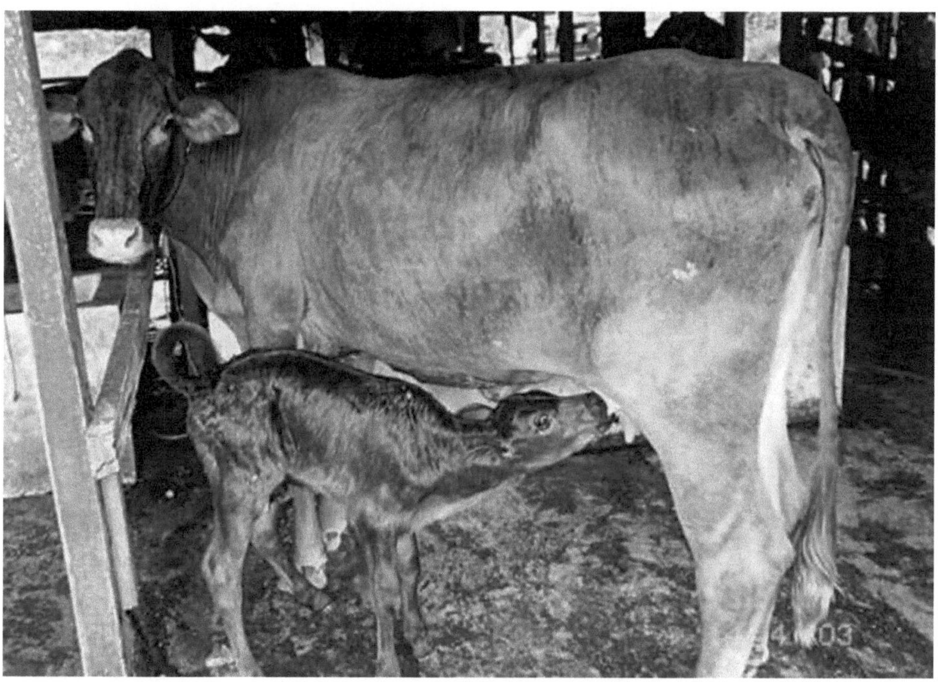

This cow requires the presence of a calf for milk let-down (Malaysia).

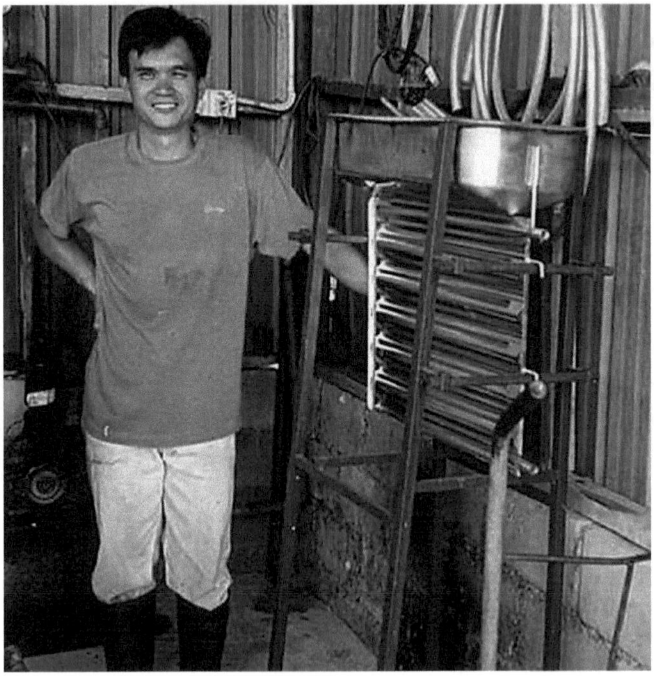

A simple plate cooler for rapidly cooling milk (Malaysia).

Very poor cleaning of milk cans; note the flies (Pakistan).

Lengthy soaking of milk clusters is not a good sterilising practice (Malaysia).

Excellent storage of milking machine equipment following use (Thailand).

Good milking hygiene practice, with separate cleaning towels for each cow (Thailand).

A simple practice such as ensuring hot water in the dairy can improve milk quality (Indonesia).

A simple gas water heater for ensuring hot water for cleaning empty milk cans at a milk collection centre (Indonesia).

Adding value to milk through making mozzarella cheese (Indonesia).

Maturing the mozzarella cheese in a cool room for 10 days (Indonesia).

Ice will not chill the milk sufficiently during power shortages at this milk collection centre (Indonesia).

Contamination of raw milk with dirt (Indonesia).

Appendix 1: Temperature Humidity Index

The following table presents the Temperature Humidity Index, calculated from temperature (in degrees Fahrenheit or Centigrade) and relative humidity (%), highlighting its potential effects on cow heat stress and hence performance.

Temperature		Relative humidity																				
Deg F	Deg C	0	5	10	15	20	25	30	35	40	45	50	55	60	65	70	75	80	85	90	95	100
72	22.2																				72	72
73	22.8																		72	72	73	73
74	23.3																72	72	73	73	74	74
75	23.9														72	72	73	73	74	74	75	75
76	24.4												72	72	73	73	74	74	75	75	76	76
77	25.0											72	72	73	73	74	74	75	75	76	76	77
78	25.6										72	73	73	74	74	75	75	76	76	77	77	77
79	26.1									72	73	73	74	74	75	76	76	77	77	78	78	79
80	26.7							72	72	73	73	74	74	75	76	76	77	78	78	79	79	80
81	27.2						72	72	73	73	74	[illegible]	75	76	77	77	78	78	79	80	80	81
82	27.8						72	73	73	74	75	75	76	77	77	78	79	79	80	81	81	82
83	28.3					72	73	73	74	74	75	76	77	78	78	79	80	80	81	82	82	83
84	28.9				72	73	73	74	75	75	76	77	78	78	79	80	80	81	82	83	83	84
85	29.4			72	72	73	74	75	75	76	77	78	79	79	80	81	81	82	83	84	84	85
86	30.0			72	73	74	74	75	76	77	78	78	79	80	81	81	82	83	84	84	85	86
87	30.6		72	73	73	74	75	76	77	77	78	79	80	81	81	82	83	86	85	85	86	87
88	31.1	72	72	73	74	75	75	76	77	78	79	80	81	81	82	83	84	85	86	86	87	88
89	31.7	72	73	74	75	76	76	77	78	79	80	81	82	83	84	85	86	86	87	88	89	89
90	32.2	72	73	74	75	76	77	78	79	79	80	81	82	83	84	85	86	86	87	88	89	90
91	32.8	73	74	75	76	76	77	78	79	80	81	82	83	84	85	86	86	87	88	89	90	91
92	33.3	73	74	75	76′	77	78	79	80	81	82	83	84	85	85	86	87	88	89	90	91	92
93	33.9	74	75	76	77	78	79	80	80	81	82	83	84	85	86	87	88	89	90	91	92	93
94	34.4	74	75	76	77	78	79	80	81	82	83	84	85	86	87	88	89	90	91	92	93	94
95	35.0	75	76	77	78	79	80	81	82	83	84	85	86	87	88	89	90	91	92	93	94	95
96	35.6	75	76	77	78	79	80	81	82	83	85	86	87	88	89	90	91	92	93	94	95	96
97	36.1	76	76	78	79	80	81	82	83	84	85	86	87	88	89	91	92	93	94	95	96	97
98	36.7	76	77	78	79	80	82	83	84	85	86	87	88	89	90	91	93	94	95	96	97	98
99	37.2	76	77	79	80	81	82	83	84	85	87	88	89	90	91	92	93	94	96	97	98	99
100	37.3	77	78	79	80	82	83	84	85	86	87	88	90	91	92	93	94	95	97	98	99	
101	38.3	77	79	80	81	82	83	86	86	87	88	89	90	92	93	96	95	96	97	99		
102	38.9	78	79	80	81	83	86	85	86	87	89	90	91	92	96	95	96	97	96			
103	39.6	78	79	81	82	83	86	86	87	88	89	91	92	94	95	96	97					
104	40.0	79	80	81	82	86	85	86	88	88	90	91	93	94	95	96						
105	40.6	79	80	82	83	86	86	87	88	89	91	92	93	96	96	97						
106	41.1	80	81	82	86	85	86	88	89	90	91	93	94	95	97	98						
107	41.7	80	81	83	86	85	87	88	89	91	92	94	95	96	98							
108	42.2	81	82	83	85	86	87	89	90	92	93	94	96	97								
109	42.3	81	82	86	85	87	88	89	91	92	94	95	96	98								
110	43.3	81	83	86	86	87	88	90	91	92	94	96	97									
111	43.9	82	83	85	86	88	89	91	93	94	95	96	98									
112	44.4	82	86	85	87	88	90	91	94	94	96	97										
113	45.0	83	86	86	87	89	90	92	95	95	96	96										
114	45.4	83	85	86	88	89	91	92	94	96	97											
115	46.1	86	85	87	88	90	91	94	95	96	98											
116	46.7	86	86	87	89	90	92	94	95	97												
117	47.2	85	86	88	89	91	93	94	96	98												
118	47.3	85	87	88	90	92	93	95	97													
119	48.3	85	87	89	90	92	94	96	97													
120	48.9	86	88	89	91	93	94	96	98													
121	49.4	86	88	90	92	93	95	97														

Five comfort zones for milking cows

A <72 No stress
B 72–78 Mild stress
C 78–89 Severe stress
D 89–98 Very severe stress
E >98 Dead cows

Appendix 2: Abbreviations and conversion of units of measurements

1. Abbreviations

k	kilo or thousands
M	mega or millions
mm	millimetre
cm	centimetre
m	metre
ha	hectare
mL	millilitre
L	litre
J	joule
MJ	megajoule
min	minute
hr	hour
yr	year
mg	milligram
g	gram
kg	kilogram
t	tonne
lb	pound
ft	foot
$	dollar
c	cent
<	less than
>	greater than

2. Conversion of Imperial units to metric units

Length:	1 inch = 25.4 mm
	1 foot = 30.5 cm

	1 yard = 0.91 m
	1 mile = 1.61 km
Volume:	1 cu ft = 0.028 m^3
	1 pint = 0.57 L
	1 gallon = 4.54 L
	1 bushel = 36.4 L
	1 acre foot = 1.23 ML
Area:	1 acre = 0.40 ha
	1 sq mile = 2.59 sq km
Weight:	1 ounce = 28.3 g
	1 pound = 0.454 kg
	1 hundred weight = 50.8 kg
	1 long ton = 1017 kg (2240 lb)
Energy:	1 calorie = 4.19 joules
Density:	1 lb/ft^3 = 0.063 kg/m^3
Rate:	1 gallon/acre = 11.23 l/ha
	1 pound/acre = 1.12 kg/ha
	1 gallon/ton = 4.17 l/tonne
Pressure:	1 pound/sq in (psi) = 1.45 kPa (kilopascals)
Yield:	1 lb/ac = 1.12 kg/ha
Temperature:	1°F = ((9/5) × C) + 32
	1 degree F is equivalent to 0.56 degrees C
	50°F = 10.0°C
	60°F = 15.6°C
	70°F = 21.1°C
	80°F = 26.7°C
	90°F = 32.2°C
	100°F = 37.8°C
	110°F = 43.3°C

3. Conversion of US units to metric units

Volume:	1 gallon = 3.79 L
	1 bushel = 35.2 L
Weight:	1 hundred weight = 45.4 kg
	1 short ton = 907 kg (2000 lb)
Milk prices:	$10/hundred weight = 22.0 c/L
Forage maize yields @ 30% DM:	
	25 ton fresh weight/acre = 16.8 t DM/ha
Food energy:	1% unit TDN = 0.185 MJ/kg DM of metabolisable energy
	30% TDN = 3.7 MJ/kg DM of ME
	40% TDN = 5.5 MJ/kg DM of ME
	50% TDN = 6.4 MJ/kg DM of ME
	60% TDN = 7.4 MJ/kg DM of ME

70% TDN = 8.3 MJ/kg DM of ME
80% TDN = 9.2 MJ/kg DM of ME
1 MCal/lb = 9.22 MJ/kg
1 MCal/kg = 4.19 MJ/kg

4. Conversion of other specific country units to metric units

Most countries now use the metric units of measurement, but certain countries have their own historical units, which are still used by farmers and advisers.

China

Length:	1 chi = 33 cm
	1 li = 500 m
Volume:	1 gongsheng = 1 L
Weight:	1 jin = 500 g

Thailand

Length:	1 nui = 2.1 cm
	1 kheup = 25 cm
	1 sawk = 50 cm
	1 waa = 2 m
	1 sen = 40 cm
	1 yoht = 16 km
Weight:	1 baht = 15 g
	1 tamleung = 60 g
	1 chang = 1.2 kg
	1 haap = 60 kg
Area:	1 sq waa = 4 sq m
	1 ngaan = 400 sq m
	1 rai = 1.6 ha

Appendix 3: Expectation and evaluation forms for workshop

IMPROVED HERD MANAGEMENT FOR HIGH GRADE DAIRY STOCK

Expectations of workshop

Location: ....

1. Name: ....
2. Address: ....
3. Position held (farmer, dairy cooperative staff, milk collection centre staff, government adviser):

   ....

4. How many milking cows do you have?

   ....

5. What is your total number of dairy stock (calves, heifers, cows, bulls)?

   ....

6. How many acres of land do you have growing forages?

   ....

7. How many litres of milk each day do all your milking cows produce (on average)?

   ....

8. What topics would you like to learn about in this workshop?

   a) ....

   b) ....

   c) ....

Please answer the following questions with a Yes or No

	Yes/No
9. Should your country be importing high grade dairy heifers from other countries?	
10. Do you think government staff are aware of the need for improved management of these animals?	
11. Do you think farmers are aware of the need for improved management of these animals?	
12. Should farmers be selected on the basis of their herd management skills?	
13. Do many farmers have the skills to manage imported heifers to achieve high milk yields and fertility?	

IMPROVED HERD MANAGEMENT FOR HIGH GRADE DAIRY STOCK

Evaluation of workshop

Location: ....

Participant's name: ....

1. **Expectations:**
   What were your expectations of the workshop? Please list:

2. **Outcome:**
   What knowledge have you gained from this workshop?

3. **Relevance of training:**
   Please describe how this training will be of use to your work.

4. **Program delivery:**
   Please tick the appropriate space to indicate your views on the way the workshop has been delivered.

	Not enough	About right	Too much
Overall program			
Lectures and/or formal instruction			
Discussion			
Visits on site/fieldwork			
Reading matter provided			

5. **Services:**
   How do you rate the services provided for you? (Please tick)

Please tick	Excellent	Good	Fair	Not good
Training/trainers				
Training location				
Other				

   Other ....

6. **Other comments:** ....

   ....

7. **Overall assessment:**
   How do you rate this program in terms of its relevance to your role in the dairy industry? (Please tick)

	Excellent	Good	Fair	Not good
Personal relevance to you				

8. **What are the weaknesses of the workshop?**

....

....

....

9. **What improvements can be made for future workshops?**

....

....

....

10. **List the most important messages/information that you found most useful to you.**

....

....

....

11. **List the least useful messages/information that you found least useful to you.**

....

....

....

Please rank the following questions for their importance to you (1 to 5), where 1 is low/ not much and 5 is high/a lot.

	Score
12. How do you rate farm visit?	
13. How do you rate small groups and reporting back sessions?	
14. How do you rate overhead presentations?	
15. How do you rate importance of improved herd management skills in your job?	
16. How much have you improved your knowledge of herd management skills?	
17. How well will you be able to apply knowledge to farmer situations?	
18. When should you do a refresher course? Please score 1 for 3 months; 2 for 12 months; 3 for 2 years; 4 for never	

Thank you for your participation in this workshop

Appendix 4: Indonesian dairy small holder pamphlet

Tips For Proper Managing of Dairy Cows On Indonesian Smallholder Farms

(Edition 2)

All dairy cows have one thing in common, their milk production and quality are directly related to the way they are looked after, and the way they are milked. Caring and gentle treatment directly affects a cow's well-being, improving its milk production and general health.

A. Housing for Dairy Cows

To ensure that the cows are in a proper environment resulting in higher milk production, the animals must be in proper shelters;

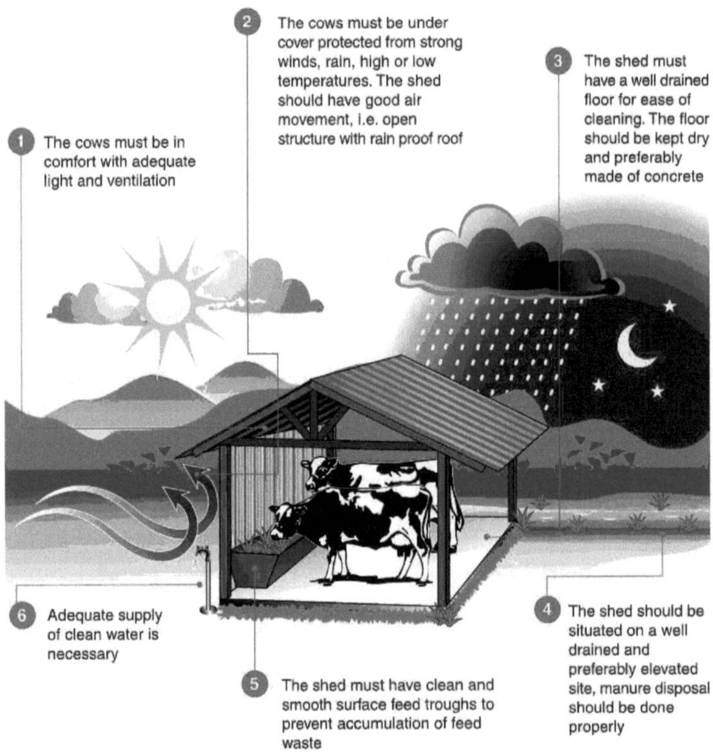

Tips For Proper Managing of Dairy Cows On Indonesian Smallholder Farms

B. Livestock Health and Hygiene

A good dairy farmer should be capable of detecting the early symptoms of ill-health in dairy cattle and to take steps to eliminate the sources of disease on the farm, e.g. contaminated water supply, infected buildings or roughages, etc. Other important factors are:

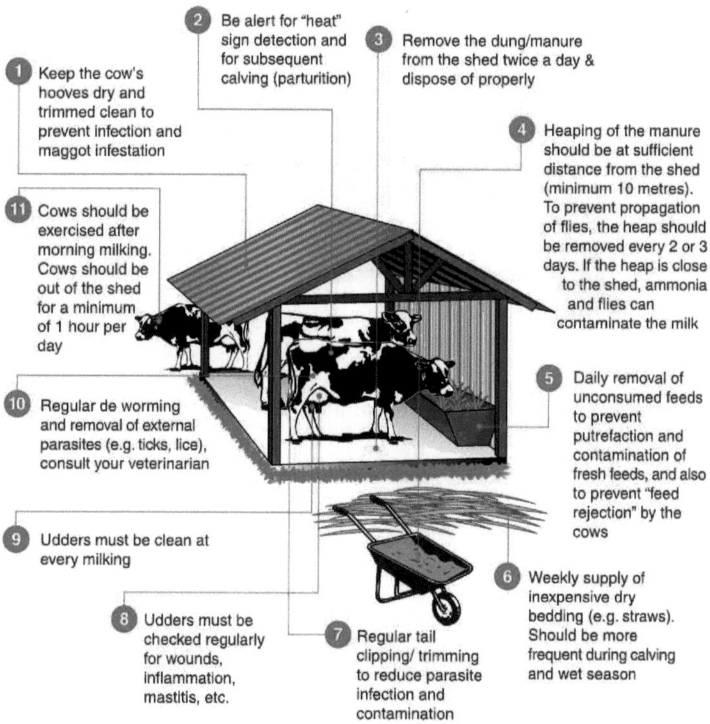

Tips For Proper Managing of Dairy Cows On Indonesian Smallholder Farms

C. Recommended Feeding

Follow the recommendations from the dairy training centre, i.e. a cow must eat the right quantity and the right quality of green roughage and concentrates, depending on the stage of development (calf, grower, heifer, pregnant, dry cow, bull) and milk production level (higher milk production needs more feeds) to maintain satisfactory milk production and ensure cow gets back in calf within 100 – 150 days of calving. Ideally milking cows should be fed 40 – 50 Kg fresh, high quality forage each day plus1 kg quality formulated concentrate per 1 – 2 L milk. Wet or dry (ampas tahu, onggok, rice bran) by products can substitute for some of the concentrates so long as they.

D. Dairy Farmer Personal Hygiene

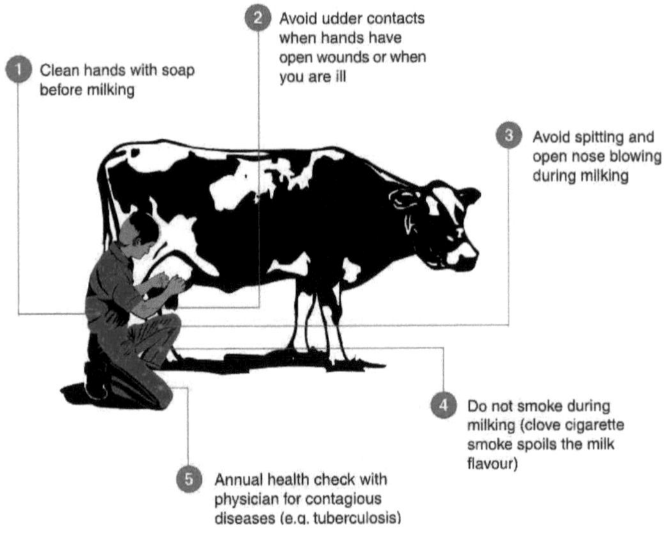

Tips For Proper Managing of Dairy Cows On Indonesian Smallholder Farms

E. Recommended Milking and Milk Handling

1. Always use clean milking made from stainless steel or alumunium after milking utensils
2. Ensure availability of adequate supply of clean water
3. Reduce sudden noise levels when milking. A disturbed cow may not "let down" her milk completely if she is excited, frightened, stressed or in pain
4. Morning and afternoon milking at same times each day
5. Teast to be wiped with a dry and clean cloth or paper towel before and after milking. If udder is very dirty, it should be washed with clean water
6. All quarters (udders) should be completely emptied of milk to reduce mastitis risk
7. Feed small amounts of concentrates during milking
8. Use only recommended and safe detergents in hot water cleaning
9. Milk to be kept as cool as possible to minimize bacterial growth, deliver to TPS/KUD quickly, avoid leaving the sun
10. Do not let morning milk over to afternoon or afternoon milk to next morning

Important:
IF COWS SHOW SIGNS OF ILL-HEALTH, CONTACT IMMEDIATELY THE DAIRY CENTRE OR VETERINARIAN ON DUTY.

All these recommendations will ensure that the dairy cows will remain healthy, thus increase milk production and improve milk quality which result in higher income.

State Government Victoria AUSTRALIA · Queensland Goverment Department of Primary Industries · GKSI · DeLaval · Tetra Pak More than the package™

Glossary

Terms in this glossary are defined in the context of their use in this book.

acid detergent fibre (ADF): the less digestible or indigestible parts of the fibre; that is, the cellulose and lignin only.

acidosis: an excessive increase in rumen acid caused by feeding too much grain or other starchy feeds or by introducing them into the diet too quickly.

ad lib (or ***ad libitum***)**:** fed to appetite.

AI: artificial insemination.

AQIS: Australian Quarantine Inspection Service.

Australian Friesian Sahiwal (AFS): a tropically adapted dairy breed developed in Australia, based on a Friesian and Sahiwal cross.

Australian Milking Zebu (AMZ): a tropically adapted dairy breed developed in Australia based on a Jersey and Red Sindhi cross.

BMP: best management practice.

body condition: the energy stored in body reserves by cows, predominantly as fat.

cfu: colony forming unit.

CIP: cleaning in place.

CMT: California mastitis test.

colony farming: a term used in Indonesia to describe a system of dairy farming where many small holders house their stock in large sheds but still maintain independent feeding and herd management.

conception rate: the proportion of the total number of services or inseminations that result in pregnancy.

comfort zone: the range of air temperature when there is no measurable fluctuation in physiological processes of cattle.

condition score: an objective visual assessment of a cow's body condition on a scale of 1 (emaciated) to 8 (obese).

cost of production (COP): the summation of variable costs, cash overheads and imputed overhead costs for producing milk from the dairy enterprise.

crude fibre (CF): a measure of fibre in the diet that is now considered unacceptable as it does not always take into account all of the constituents that make up the fibre component of a feed; it measures only the alkali-soluble lignin and the cellulose.

crude protein (CP): a rough measure of all the protein in the diet (NPN + RDP + UDP); it assumes (incorrectly) that all the nitrogen in a feed comes from protein.

depreciation: the loss in value of capital items as they get used or become older.

digestibility: the proportion of the dry matter in a feed that gets digested; it is the difference between what is eaten and what comes out as manure.

dry matter (DM): the proportion of any feed remaining after all the water has been taken out.

energy: the part of a feed that is used as 'fuel' in carrying out the cow's bodily functions.

enterprise: a farming activity or the production of a particular commodity or group of related commodities.

fibre: the cell wall, or structural material, in a plant made up of (among other things) cellulose, hemicellulose and lignin.

genotype by environment interaction(G × E): the relative performance of different genotypes of cattle depend on the environment under which they are managed.

imputed: the same as unpaid or book value.

key performance indicators (KPIs): measures of dairy farm performance to provide realistic targets following improvements in feeding, herd and farm management.

metabolisable energy (ME): the amount of energy provided by a feed after deducting energy lost to faeces, urine, heat, and gas production; it is the energy available to be used by the cow for her metabolic activities.

methylene blue reductase test (MBRT): a laboratory test for quantifying levels of bacterial contamination in raw milk.

MJ ME/kg DM: megajoules of metabolisable energy per kilogram of dry matter.

multiple ovulation and embryo transfer (MOET): the practice of collecting embryos to fertilise *in vitro* then transport them to other regions for implanting into donor cattle.

N: nitrogen.

neutral detergent fibre (NDF): a measure of all the fibre (hemicellulose, lignin and cellulose) in a feed; it indicates how bulky the feed is.

non-protein nitrogen (NPN): not actually protein, but simple nitrogen; however, microbes can make protein from simple nitrogen if enough energy (as carbohydrates) is available in the rumen at the same time.

pH: a measure of acidity or alkalinity on a scale from 1 (extremely acid) to 14 (extremely alkaline).

protein: the material that makes up most of the cow's body (muscles, skin, organs, blood); it also is part of milk.

quality: in relation to feeds, it is an indication of the level of energy and digestibility. In relation to milk, it refers to the level of various contaminants in milk, such as bacterial, chemical or any other adulterations that can be detected.

retained foetal membrane (RFM): membranes from newly born calves still inside the cow following birth.

rolling herd average: the average milk yield of all the lactating cows in the milking herd

rumen degradable protein (RDP): the portion of protein in the diet that is digested and used by the microbes in the rumen to build themselves, if enough energy (carbohydrates) is available at the same time.

rumen undegradable protein: *see* undegradable dietary protein.

Rp: Indonesian rupiahs.

Rs: Sri Lankan rupees.

SHD: small holder dairy.

Solids-not-fat (SNF): an alternative measure of milk composition to milk protein. SNF contains milk protein, lactose and minerals, allowing the percentage of milk protein to be calculated as (SNF% – 5.4).

somatic cell count (SCC): a measure of number of white blood cells in milk used to quantify extent of mastitis.

specific gravity: a measure of weight relative to volume; for milk, this can vary from 1.024 to 1.032. When quantifying milk yields, Western dairy specialists tend to use volume (L/day) because this is measured by most milking machines. In contrast, Asian dairy specialists tend to use weight (kg/day) because this is measured by most hand milking operators. Interchanging the volume and weight of milk can then introduce an error of 2–3% in milk yields.

submission rate: the proportion of the herd inseminated at least once in a given period of time (e.g. the first 10, 21, 24 or 30 days of mating).

supplement: a feed or product added to the cow's diet to increase the intake of some dietary component, such as energy, protein, fibre, vitamins or minerals.

Temperature Humidity Index (THI): a system for quantifying heat stress based on temperature and humidity. The higher the index, the greater the discomfort, and this occurs at lower temperatures for higher humidities.

total digestible nutrients (TDN): a system of describing dietary energy based on proximate analyses (ash, nitrogen, ether extract and crude fibre). The formula uses crude protein, crude fibre, ether extract and nitrogen-free extract.

total plate count (TPC): a measure of bacterial contamination of raw milk in millions of bacterial colony forming units (cfu) per mL of milk.

total solids (TS) or total dissolved solids (TDS): a measure of milk composition expressed as a percentage of total milk solids (milk fat, milk protein, lactose and minerals).

undegradable dietary protein (UDP): any protein in the diet that passes through the rumen without breaking down and is digested in the abomasum and small intestine. Also known as bypass protein.

References and further reading

Chantalakhana, C., Korpraditsakul, R., Skunmun, P. and Poondusit, T. (1999). Environmental conditions and resource management in small holder dairy farms in Thailand II. Effects of dairy wastes on water and soil. *Asian-Aus. J. Anim. Sci.* **12**, 220–5.

Chantalakhana, C. and Skunmun, P. (2002). *Sustainable Smallholder Animal Systems in the Tropics.* Kasetsart University Press, Bangkok.

Cook, B.G., Pengelly, B.C., Brown, S.D., Donnelly, J.L., Eagles, D.A., Franco, M.A., Hanson, J., Mullen, B.F., Partridge, I.J., Peters, M. and Schultze-Kraft, R. (2005). *Tropical Forages: An Interactive Selection Tool.* [CD-ROM], CSIRO, DPI&F(Qld), CIAT and ILRI, Brisbane, Australia. <http://www.tropicalforages.info>

DAFF (2006). *Australian Standards for the Export of Livestock.* Version 2, Nov 2006. <http://www.daff.gov.au/livestockexportstandards>

Edge, M.K., Barnett, J.L., Phillips, L. and Chamberlain, P. (2006). *Animal Welfare Reference Document for Dairy Production.* DPI Victoria and Dairy Australia, Melbourne.

FAO (2005a). *Grassland Species.* Detailed descriptions and photos of more than 600 grassland species. FAO, Rome. http://www.fao.org/ag/AGP/AGPC/doc/GBASE/Default.htm.

FAO (2005b). *Country Pasture Forage Profiles.* Detailed descriptions and photos of climate, livestock and forage resources. FAO, Rome. http://www.fao.org/ag/AGP/AGPC/doc/Counprof/regions/index.htm.

Freeman, M. (1993). *Your Heifers in the Balance.* DPIF, Tasmania.

Hooten, N. (2008). Dairy development for the resource poor. Lessons for policy and planning strategies. In *Developing an Asian Regional Strategy for Sustainable Smallholder Dairy Development.* Proceedings of an FAO/APHCA/CFC funded workshop, Feb 2008, Chiang Mai, Thailand. pp. 48–51, 2008. http://www.fao.org/docrep/011/i0281e/i0281e00.htm.

Horne, P.M. and Stur, W.W. (1999). *Developing Forage Technologies for Smallholder Farmers: How to Select the Best Varieties to Offer Farmers in Southeast Asia.* ACIAR Monograph 62. ACIAR, Canberra. http://aciar.gov.au/publication/mn062/.

Humphries, L.R. (1999). Forage utilisation. In *Smallholder Dairying in the Tropics.* (Eds L. Falvey and C. Chantalakhana) pp. 185–201. International Livestock Research Institute, Nairobi, Kenya.

Ibrahim, M. N. (1988). *Feeding Tables: A Practical Guide.* p. 71. E.T.I. Division, Animal Production and Health, Kandy, Sri Lanka.

Ibrahim, M.N., Wassing, G.J., Widodo, M.W. and de Jong, R. (1992). Dairy cattle development in East Java. A decade of progress and future prospectives. In *Livestock and Feed Developments in the Tropics.* (Eds M.N. Ibrahim, R. de Jong, J. van Bruchem and H. Purnomo), Malang, Indonesia. p. 23.

McDowell, R.E. (1994). *Dairying with Improved Breeds in Warm Climates.* Kinnic Publishers, Raleigh, NC, USA.

Makeham, J.P. and Malcolm, L.R. (1986). *The Economics of Tropical Farm Management.* Cambridge University Press, Cambridge, UK.

MLA (2006). *Recognising Excessive Heat Load in Feedlot Cattle: Tips and Tools.* Meat and Livestock Australia, Sydney.

MLA (2009). *Advisers Guide to Good Dairy Herd Management in the Tropics: A Manual for Dairy Advisers, Technical Staff and Other Dairy Specialists.* Meat and Livestock Australia, Sydney.

Moran, J.B. (2002). *Calf Rearing: A Practical Guide.* 2nd edn. Landlinks Press, Melbourne.

Moran, J.B. (2005). *Tropical Dairy Farming: Feeding Management For Small Holder Dairy Farmers in the Humid Tropics.* Landlinks Press, Melbourne. <http://www.publish.csiro.au/nid/197/issue/3363.htm>

Moran, J. (2008). 'Increasing milk supplies in Central Java through importing high quality dairy stock'. Consultancy report to GTZ. DPIV report, Kyabram, Victoria.

Moran, J.B. (2009a). *Business Management for Tropical Dairy Farmers.* CSIRO Publishing, Melbourne. <http://www.publish.csiro.au/nid/220/issue/5522.htm>

Moran, J. (2009b). Key performance indicator's to diagnose poor farm performance and profitability of smallholder dairy farmers in Asia. *Asian-Aus. J. Anim. Sci.* **22**, 1709–17.

Moran, J.B. (2012). *Rearing Young Stock on Tropical Dairy Farms in Asia.* CSIRO Publishing, Melbourne.

Moran, J., Edwards, D. and Kharisma, A. (2006). 'Value adding Indonesia's dairy industry: developing cottage industries in East Java'. Report for Australian Business Volunteers, May 2006.

Moran, J.B. and Mickan, F.J. (2004). Increasing intakes of quality forages by cows on Indonesian small holder dairy farms. *Proceedings of the 11th Animal Science Congress, Asian-Australasian Association of Animal Production Societies, Kuala Lumpur, Malaysia,* Vol. 2, pp. 7–10.

Moran, J.B. and Tranter, B. (2004). *Reproductive Management of Small Holder Farmers in Vietnam.* F & N Vietnam Foods and DPI Kyabram, Victoria.

Morton, J., Larcombe, M. and Little, S. (2003). *The InCalf Book for Dairy Farmers.* Dairy Australia, Melbourne.

Muia, J.M., Tamminga, S., Mbugua, P.N. and Kariuki, J.N. (2000). The nutritive value of Napier grass (*Pennisetum purpureum*) and its potential for milk production with or without supplementation: a review. *Tropical Science* **40**, 109–31.

National Research Council (1989). *Nutrient Requirements of Dairy Cattle* (6th edn). National Academic Press, Washington DC, USA

Sprecher, D.J., Hostetler, D.E. and Kaneene, J.B. (1997). A lameness scoring system that uses posture and gait to predict dairy cattle reproductive performance. *Theriogenology* **47**, 117–87.

STOAS (1999). Fertilisation of pasture. In *Modern Dairy Farming in Tropical and Subtropical Regions*. Vol. 1. pp. 69–71. STOAS, Wageningen, Netherlands.

Thirapatskun, T. (1999). Mastitis management. In *Smallholder Dairying in the Tropics*. (Eds L. Falvey and C. Chantalakhana) pp. 299–321. International Livestock Research Institute, Nairobi, Kenya.

Tibbo, K., Wiener, G. and Fielding, D. (1994). A review of the performance of the Jersey breed of cattle and its crosses in the tropics in relation to the Friesian or Holstein and indigenous breeds. *Animal Breeding Abstracts* **62** (10), 719–757.

West, J.W., Mullinix, B.G. and Bernard, J.K. (2003). Effects of hot humid weather on milk temperature, dry matter intake and milk yield of lactating dairy cows. *Journal of Dairy Science* **86**, 232–242.

Index

9 780643 107403

Printed by Libri Plureos GmbH in Hamburg,
Germany